# The Innovation Game

# The Innovation Game

## A New Approach to Innovation Management and R&D

by

**Armelle Le Corre**
*Ballard Power Systems Corp., Product Development*

*and*

**Gerald Mischke**
*DaimlerChrysler Corp., Research & Technology Project Controlling*

 Springer

Library of Congress Cataloging-in-Publication Data

A C.I.P. Catalogue record for this book is available
from the Library of Congress.

ISBN 0-387-23794-1     e-ISBN 0-387-23843-3        Printed on acid-free paper.

Printed in the United States of America.

9 8 7 6 5 4 3 2 1       SPIN 11344315

springeronline.com

# Contents:

# List of questions to be answered:

# List of inno- and investment management rules derived:

# List of definitions:

# List of lemmas:

# List of figures:

# Introduction and abstract

Innovation management is all about **bringing new life and spirit into old economies and paralyzed markets.** There are numerous conflicts to be settled in this process, all of which have at least one common reason. This main reason is, that

- **the conflicting parties in general do not see the common interest** behind their more or less well founded particular interests at the moment.

Some examples for the main conflict areas in this struggle are

1) **investors** not understanding nor evaluating properly the market chances and risks together with the respective technology options and pitfalls of an innovation project.

2) **R&D-managers** and accountants trying to optimize their schedules and budgets without paying proper respect to the individual project's options to cope with those restrictions.

3) **individual projects** struggling for budgets and time to completion without paying proper respect on which other opportunity might be better for the markets and the company as a whole.

We firmly did and do believe, that there is a common interest in all these conflicts, cause, just as Schumpeter pointed out in his fundamental theory on innovation [1],

- **innovation, once it is achieved, is a winning game for almost anybody taking part!**

Strengthening and alleviating innovations had been the objective, once we started our endeavor to look into the details of this process. Very early we learned, that **cross-functional thinking and knowledge is the key** to understand it in more detail and, most of all, that **a much more detailed understanding and modeling of the innovation-process is mandatory** to achieve our objective.

In doing this, we were able to set up a general innovation-process model, the inno-gem. It describes the innovation process on 3 complementary layers or views, each of which is addressing a typical problem class between conflicting interests. These layers are

1)  the **macroscopic view** - addressing the design and the capacity optimization rules for complete inno-pipes and/or enterprises.

2)  the **mesoscopic view** - addressing in particular the management and the control problems of inno-pipes including the design of appropriate quality gate approaches.

3)  the **microscopic view** - describing the effects of control and design decisions taken on the other two layers on the success chances of individual inno-projects and, vice versa, how different strategies, methods and technologies applied there do influence complete inno-pipes and/or firms.

In the first 4 chapters we present and discuss our inno-process model leading us directly to a new and **sound foundation of knowledge management for R&D** in chapter 4. Additionally we show in the $5^{th}$ chapter, how our **theory does fit to the principles, methods and instruments of classical finance and investment management**. As a result, we do **propose a new improved inno-accounting principle,** much better suited to judge and manage inno- and R&D-investments than the ones used so far.

Therefor we do hope that this book, our model and our theory will give management a much better comprehension of the innovation-process and its peculiarities, than it has been possible before. We firmly do believe, that this better understanding does help to make any innovation endeavor to the very winning game it always should have been for anybody involved.

# How to read and use this book

This book is intended to be **used as a handbook for innovation management, very much in a way technical handbooks are structured and used.** Thus we do provide all the information necessary to properly execute and apply our proposed model, its rules and advises for an improved innovation management. This information is comprised of the following 4 basic elements:

a) All the **definitions, lemmas, discussions & proofs** (in grey boxes) needed to most explicitly render the tools, notions and principles necessary to understand and apply in a real life environment the proposed theory, its rules and its methods for a better innovation management.

b) **Questions** (in light gery boxes) and the respective subsequent paragraphs to answer and explain the problems addressed.

c) **Rules** (in dark grey boxes) to give a rather solution driven reader quite compact and direct advice on how to implement our proposed principles for a better inno-management in his environment.

d) **Intermediate and final summaries** at the end of each chapter to act as a compilation and as a quick reference to the solutions and the respective solution principles for the covered key problem area to be addressed by innovation management.

In order to ease and structure the use of these four kinds of information outlined in our handbook, **we assumed 4 basic types of readers to be supported.** Each of these 4 types of readers is assumed to have his own specific approach to innovation management and thus to our handbook. These types of readers assumed and our advises for them on how to use this book respectively are the following:

1) **The "scientist"** - who is assumed to desire an equally **comprehensive and deep understanding** of inno-management, of our theory and model and its application options and limitations as well. We would advise him

- to **study the book from start to end or to concentrate on specific chapters** once he is only interested in specific problem domains.

- to **read the proofs & discussions** carefully to be able **to judge the quality** and the limitations of our definitions, lemmas, rules and arguments properly.

- to **take the intermediate summaries and the final summary as a short wrap up** of our theory and model.

2) **The "analyst"** - who is assumed to be quite **familiar with innovation management** and the corresponding literature and approaches. We thus think that his main interest is in finding out details and differences of our approach compared to others. We would advise him to

- find the appropriate paragraphs in our book by **scanning the list of contents or** even **the list of definitions and lemmas** to directly select the topics of interest to him

- To carefully **study** the boxed proofs & discussions in order to evaluate the validity of our rules, lemmas and definitions derived

- to **take the intermediate and final summaries as a short compendium** of our approach, our model and theory and its respective implications.

3) **The "digger"** - who is assumed to look for answers and explanations for problems he does experience in his environment. We would advice him

- to **use the** list of questions to guide his search most effectively directly to the explanations and answers in our book relevant to him.

- to **read the subsequent paragraphs** until the end of the respective chapter or until the next question is reached to get the answer/explanation for the question or problem he encounters.

- to **skip the** proofs & discussions, assuming our definitions and lemmas are OK.

- to **use** the rules **in his very paragraph** as a quick answer or solution strategy to his problem.

- to **use the many references** rendered to look for additional information and to understand the answers given in more detail once this is desired or it turns out to be necessary.

4) **The "craftsman"** - who is assumed to be mainly interested in (quick) fixes for problems and/or difficulties he does experience in his environment. We would advice him

- to **use the** | **list of rules** | to directly **jump to the rule and/or advice helpful** for the problem he is confronted with.

- to **additionally use the** | **list of questions** | to directly **address some extra explanation** he might want or need.

- **use the references** rendered **for additional information** in case of questions.

- skip | **the proofs & discussions** | **and the calculus and the formalisms** used and to **rely on the verbal explanations** given, trusting in the quality of our approach and in the respective arguments used.

- **Adopt the "digger-mode" or the "analyst-mode"** once he finds himself **uneasy with some rules/explanations** given in the text.

Naturally any real reader of this book will not fit exactly to any one of these idealized categories. But we do hope, that the advice given, does help every reader to quickly find the information relevant to him, once it is in this book at all. Covering almost all aspects of innovation management, we really do hope this book is a help for quite a few people interested in this fascinating area.

# 1. THE INNOVATION PROCESS MODEL – DEFINITIONS AND LEMMAS

## 1.1 The macroscopic view - the filter paradigm or the pipeline/system layer

Before being able to set up theories and models on innovation, the terms used and their respective semantics just have to be made quite clear. Thus let us start with some necessary definitions:

*Definition 1 - the innovation $I_k$*

*Following Schumpeter [1] who first introduced the notion "innovation" to economic theory an*
   *(D 1)    innovation $I_k$ is an invention $Ts(I_k)$ having at least some market success $Ms(I_k)$.*
*Thus we can define the notion Innovation more precisely through this defining equation (D 1):*

| | | | |
|---|---|---|---|
| **(D 1)** | $I_k$ | = | *$Ts (I_k)$ & $Ms(I_k)$* |
| *with* | **Ts** | = | *Technology success* |
| *and* | **Ms** | = | *Market success of $I_k$* |

Being an innovation, $I_k$ must be new, feasible and improve at least to some extent current technology. Having a market success does mean in economical terms that, the total costs $Tc(I_k)$ to produce an innovation $I_k$ are less or equal its respective market profits $Mp(I_k)$ achieved.

*Lemma 1 - the success condition for the innovation $I_k$*

*In a more formal way we can state this lemma 1 out of the definition (D 1):*
   *(L 1)    Costs $(I_k)$ = $\Sigma$ (disc. Net-Costs $(I_k)$) = $Tc(I_k)$ < $Mp(I_k)$*
               *= $\Sigma$ (disc. Net-Profits $(I_k)$)*

This is nothing but the classical economical stability condition every product and its respective discounted cash flow (DCF) curve must fulfill along its product life cycle (see Fig. 1 ):

**Remember:** It is **the coincidence of** technological $Ts(I_k)$ **and of** market success $Ms(I_k)$ only which **does lead to an innovation !** ([i])

---

[i]   Thomas Alva Edison (1847-1931): "An invention that does not sell, I just do not want to make"

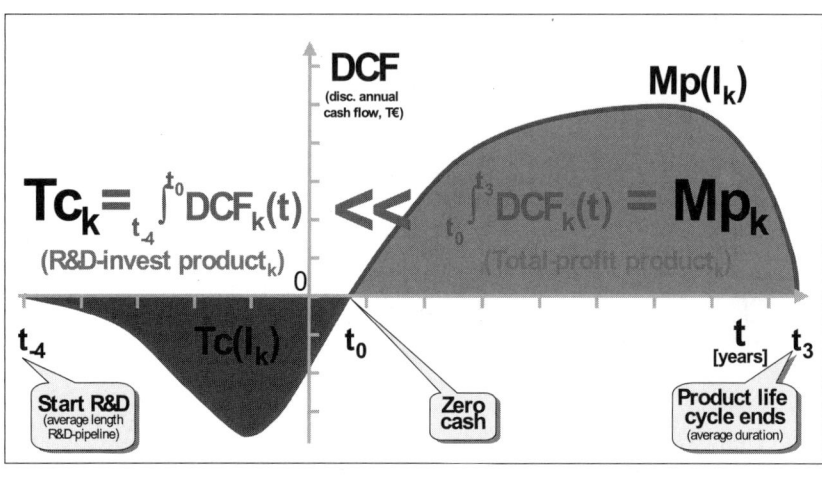

*Fig. 1: Discounted cash flow function of an innovation (product) along its life cycle*

### Definition 2 - the innovation project $Ip(I_k)$

*An innovation project $Ip(I_k)$ to come to a potential innovation $I_k$ is a sequence of N steps*

(D 2)     $Ip(I_k)$     $= \{Ip_0(I_k), .., Ip_{N-1}(I_k), Ip_N(I_k)\}$

with     $Ip_N(I_k) = Ts(I_k)$ & $Ms(I_k)$

*leading from an idea $Ip_0(I_k)$ for an innovation or a new product to its respective coinciding technology $Ts(I_k)$ and market success $Ms(I_k)$ or failure!*

Assuming this "Schumpeter- like" definition (D 2) of the innovation process, the next fundamental question is the following one:

(Q 1)     **What are the basic properties of and which forces do drive an innovation process?**

To answer these questions it is a wise policy to approach them from two sides, empirically and theoretically as well. First we would like to summarize the empirical evidence available so far:

- There is a rich abundance of empirical investigations on innovation but astonishingly there are only very few comprehensive investigations on basic properties of innovation processes and pipelines.

The one of Stevens and Burley [2] is one of the most exhaustive ones in that respect and thus recommended for reading. It clearly demonstrates (see Fig. 2) that there is very good empirical evidence for a strong statistical dependency, in general an **exponential dependency of the chance to become an innovation Ps($I_k$) from its respective state of maturity** (see Fig. 2).

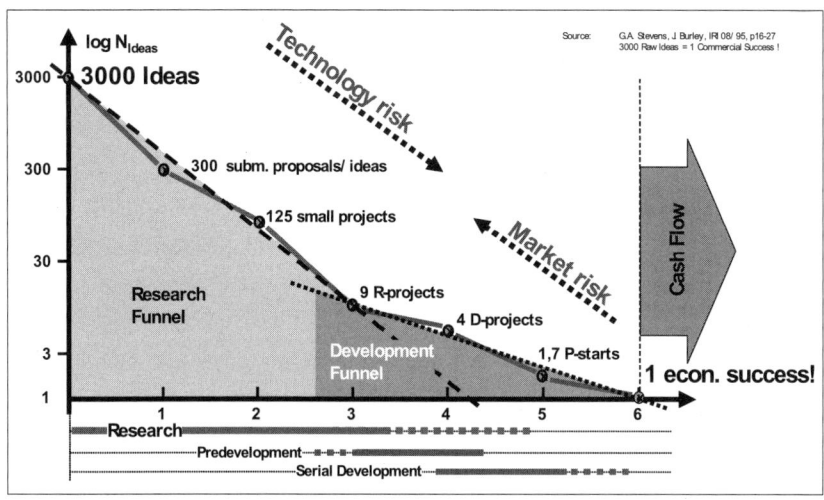

*Fig. 2: R&D-Pipeline success chance statistics[ii]*

This is strongly supported by the results of our own investigation [3] which analyzed the innovation history of more than 50 R&D projects employing more than 250M€ R&D-budget and covering a "Time to Market" (TtM-) range of up to 15 years. Here too we found a strong dependency of the chance for innovation success Ps($I_k$) from the technologically scaled "Time to Market" (TtM) of the respective innovation idea and/or project. Based on these most impressive empirical findings (see [2] and [3] for additional information) we can state the following Lemma 2 to answer question (Q 1):

*Lemma 2  -  the **TtM**-dependency of the innovation success function **Ps***

*Each innovation project **Ip($I_k$)** is a statistical filter process, where the probability **Ps(Ip($I_k$))** to become a success shows an exponential dependency of its respective market distance or maturity measured as "Time to Market" (**TtM**). Here time is always measured backwards from completion of the respective innovation, thus time is 0 at the completion of the innovation.*

$$(L\ 2)\quad Ps(Ip(I_k)) \sim exp\ (-\lambda * TtM(I_k))$$

In order to be able to prove this lemma, we need some definitions first:

---

[ii]   see G.A.Stevens, J.Burley, IRI 08/95 p. 16-27

*Definition 3 - the innovation pipeline **IP***

*An innovation pipeline **IP** is an ordered set of innovation projects **Ip(I$_k$)**, as described in chapter 1-2.1, employing a predefined set of technologies and addressing also a given set of different markets:*

$$(D\ 3) \qquad IP = \{\ Ip(I_1),........\ Ip(I_n)\ \}$$
$$\text{with} \qquad I_k = Ms(I_k)\ \&\ Ts(I_k) \qquad\qquad for \quad k = \{1....n\}$$

*Definition 4 - the success-function **Ps(IP)** of an innovation pipeline **IP***

*With definition (D 3) above it is ensured that the success-function Ps of an inno-pipe **IP** can easily be defined as the weighted average of the Ps-values of its respective inno-projects **Ip$_k$**:*

$$(D\ 4) \qquad Ps\ (IP(t_i)) = \sum_k Ps(Ip_k(t_i)) * weight(Ip_k)$$

Here $t_i$ is a time defined between 0 and the maximum TtM of IP. The success probability Ps(IP($t_i$)) of an Inno-pipeline IP observable at some stage or time-point $t_i$ is thus some statistical average of the respective individual success probabilities Ps($I_k(t_i)$) = Ps(Ip($I_k(t_i)$)) of the respective innovations $I_k$ or innovation projects Ip($I_k$) in the same stage $t_i$.

If one does not want to compare apples with oranges, one just has to sort the individual innovations corresponding to their state of maturity, the markets they are aiming at and the technologies they are employing (see also chapters 2.4, 2.6, 3.3, 3.4 and definitions (D 13) and (D 14)). Doing this, it is to some extend guaranteed that lemma 3 holds:

*Lemma 3 - the **TtM**-dependency of the innovation-pipeline success function*
**Ps(IP)**

$$(L\ 3) \qquad Ps(IP(t_i)) \sim exp(-\lambda * TtM(IP(t_i)))$$

The success probability of an innovation pipeline Ps(IP($t_i$)) at stage $t_i$ is obeying the same exponential characteristic as its underlying set of innovation projects Ip$_i$. This is demonstrated in Fig. 3, which shows a plot of a typical success function Ps (t) along a product cycle.

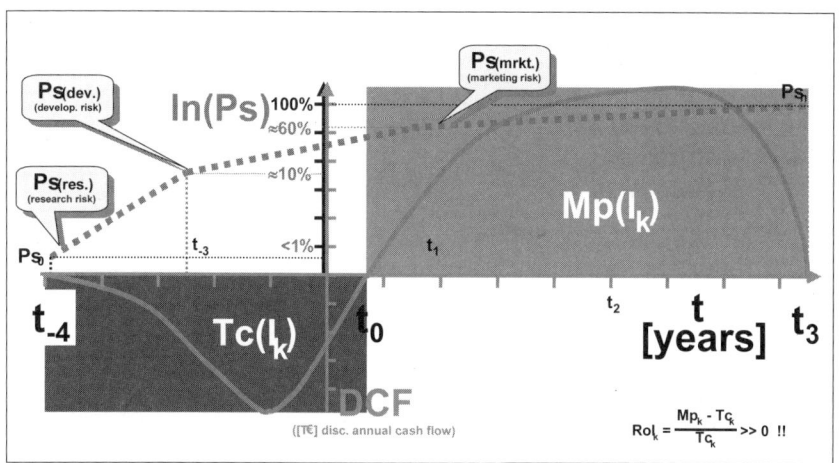

*Fig. 3: Plot of a typical success function curve Ps(t) of an inno-project or -pipeline*

**Proof of Lemma 2 and of Lemma 3:**

   To prove that, by necessity, we really do have an exponential decrease of the observed success probability Ps, we have to inspect an individual innovation project more closely. Remember that we have defined an innovation project as an ordered set of steps $Ip(I_k) = \{Ip_0(I_k), ..., Ip_{N-1}(I_k), Ip_N(I_k)\}$ leading to an innovation-success, which is equivalent to $Ip_N(I_k) = Ts(I_k)\&Ms(I_k)$.

**Proof step 1):**

   At the start $Ip_0(I_k)$ you state with some "credibility" $Ps_0$ the proposition that $I_k$ will be an innovation. Written in the terms of the probability calculus, this statement or proposition is equivalent to:

$$Ps(Ip_N(I_k)) = 1 \quad \text{and} \quad Ts(Ip_N) = Ms(Ip_N) = 1$$

**Proof step 2):**

   Thus during each innovation project $Ip(I_k)$ it must be true that at least at some stage i

$$Ps(Ip_i(I_k)) < Ps(Ip_j(I_k)) \quad \text{if } i<j$$

This just states that you have to achieve some progress towards your innovation goal, at least from some stage i on. Therefore every innovation project Ip can be ordered into a time sequence of stages $t_i$ with Lemma 4:

*Lemma 4 - the optimal step by step innovation progress*

*(L 4)*    $Ip(I_k) = \{Ip\,(t_1), ..., Ip\,(t_n)\} = \{Ip_1, ..., Ip_n\}$    *and*
          $Ps(I_p(t_i)) < Ps(I_p(t_j))$    *for i<j*

*Where $Ip(t_i)$ is the status of the innovation project $Ip(I_k)$ at a set time point $t_i$.*

These steps represent stages of progress towards an innovation. Any quality gate process like e.g. in vehicle development (e.g. the Mercedes[iii] MDS process) is nothing but a practical application of this lemma (L 4), just as the following definitions ((D 5), (D 6)) do show us:

*Definition 5 - the quality gate process $Qg(IP,t)$*

*A quality-gate process $Qg(IP,t)$ is a time, better a TtM-sequential set of n evaluations of the values of the innovation-success functions $Ps(Ip)$ of a set of m innovation-projects $IP=\{Ip_1, ..., Ip_m\}$:*

$$(D\ 5)\quad Qg(IP,t) = \{Qg_0, ... , Qg_n\}$$

Each quality gate $Qg_i$ reviews for each inno-project $Ip_j$ at a set time-point $t_i$ (e.g. corresponding to a predefined project maturity stage i) whether or not the values of its respective success-function $Ps(Ip_j,t_i)$ are within certain predefined boundaries $\varepsilon > 0$ of a given predefined threshold value $Ps_i$:

*Definition 6 - the quality gate $Qg_i(Ip,t_i)$*

$$(D\ 6)\quad Qg_i(Ip_j,t_i) \Leftrightarrow Ps_i\text{-}\varepsilon < Ps(Ip_j, t_i) < Ps_i\text{+}\varepsilon$$

*where each $Ps_i = P(Ts_i\ \&\ Ms_i)$, with $0 \leq i \leq n$, is an element of an ordered set of values:*

$$Ps_i = \in \{Ps_0, ... , Ps_n\}$$

Starting from some arbitrary start-risk or -probability $Ps_0$, the $Ps_i$ have the properties:

*Lemma 5 - the innovation progress $Ps_i$ between gates $i$ and $i+1$*

$$(L\ 5)\quad Ps_i < Ps_{i+1}\ and\ Ps_n = 1$$

---

[iii]    See e.g. presentation of Dr. Schöpf, head Mercedes development on the ATZ/MTZ-conference "Virtual Product Creation" 2004 in Stuttgart

## 1.2    The microscopic view - the logical proof tree paradigm or the project layer

**Proof step 3 – the most optimal innovation project:**

To prove this we have to introduce a new inno-project layer, for up to now, we only know that there always is a way to make a step by step progress in an innovation project and in the corresponding inno-pipe too. We still do not know whether there is an exponential growth of the chance for success Ps(t) with time. So let us assume a most optimal innovation project Ip with the following properties:

**a)** The most optimal innovation project Ip consists of just N steps

$$Ip = \{Ip\,(t_1)........Ip\,(t_N)\}$$

**b)** There are just two alternatives possible at each step and there is always only one alternative leading towards the innovation success (there is always only just one better possibility and only one optimal path to innovation success). With these properties we can apply the rules of logical calculus (theorem proving) on our Schumpeter like definition of an innovation project (**D 2**)

$$(\text{D 2}) \qquad\qquad Ip(I_k) \;=\; \bigcup_{1}^{N} Ip(t_i)$$

The corresponding model rendered from logical calculus is the N-level proof-tree of $Ip(I_k)$ shown in Fig. 4. One immediately sees from this proof tree or the equivalent decision tree that the probability of success Ps exponentially decreases with the number of steps (quality stages) necessary for achieving an innovation success.

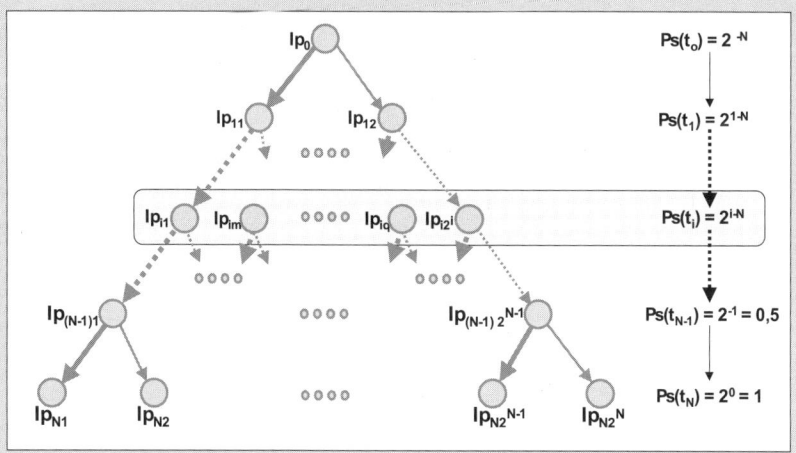

Fig. 4: Proof tree of the most optimal inno-project $Ip(I_k)$

**c)** If we assume, as indicated in Fig. 4, that each step $Ip(t_i)$ to $Ip(t_{i+1})$ in an innovation tree is equidistant and that it renders the same positive contribution to the overall success Ps,

$$\textbf{Ps(Ip}(t_{i+1}))/\textbf{Ps(Ip}(t_i)) = c > 1$$

we then see immediately, that due to $Ps(Ip(t_{i+1})) = Ps(Ip(t_i + \Delta t)) = c*Ps(Ip(t_i))$

$$\begin{aligned}\Delta Ps / \Delta t \; &= \; [Ps(Ip(t_i + \Delta t)) - Ps(Ip(t_i))] / \Delta t\\ &= (c-1) * Ps(Ip(t_i)) / \Delta t\end{aligned}$$

we can make the transition $\quad \lim\limits_{\Delta t \to 0}(\Delta Ps / \Delta t) = dPs = (c-1)Ps/dt$

and derive as a consequence $\quad \int Ps * dt = \dfrac{-1}{\lambda} Ps \quad$ with $\lambda = 1/(1-c)$

(Proof of (L 2))    $Ps(Ip) \sim \exp(-\lambda*t_i)$    **Q.E.D.**

This proves the empirically found lemma 2 for any inno-project Ip. Together with (D 3), (D 4), (D 5), (D 6), (L 4) and (L 5) this proofs Lemma 3 for inno-pipes $IP = \{Ip(I_1), ..., Ip(I_n)\}$ too:

(Proof of (L 3))    $Ps(IP(t_i)) \sim \exp(-\lambda*TtM(IP(t_i)))$    **Q.E.D.**

We may now state, that the exponential time-dependency of the innovation-success probability Ps for any inno-project or inno-pipe is a logical consequence of the (our) Schumpeter-like definition of an innovation. With these definitions, lemmas and proofs, we can fairly well describe some of the most basic properties of inno-pipes and of inno-projects too. Thus we are able to give a most decent first approach to the complete answer to question 1.

**(Q1)    "Which are the basic properties of an innovation process?"**

What we still cannot describe properly is how an inno-pipe, a set of inno-projects or an individual one do respond to some action taken (e.g. by management) or to some input given. When does it respond how on a specific input? To be able to get some more insight into this problem we have to introduce a new view or description layer – the mesoscopic view in the next chapter.

## 1.3    The mesoscopic view - the control paradigm or the management layer

To be able to give a satisfying answer on the most crucial question for R&D-management "Which forces do drive the results of an innovation process?", we have to introduce a new view or layer of description:

**The mesoscopic view or the innovation pipeline control layer.**

If one really wants to execute control on any system, one just has to know its system function or model. For an innovation pipeline we know that it behaves like an exponential filter for the innovation success probability Ps (see (L 2) showing some characteristic delay time (TtM). For systems like that, control theory renders us a system model composed of

- a **selection or a filter function element representing Ps(t) and**
- a **delay element TtM representing the pass through time of the respective filter,**

as indicated in Fig. 5 below.

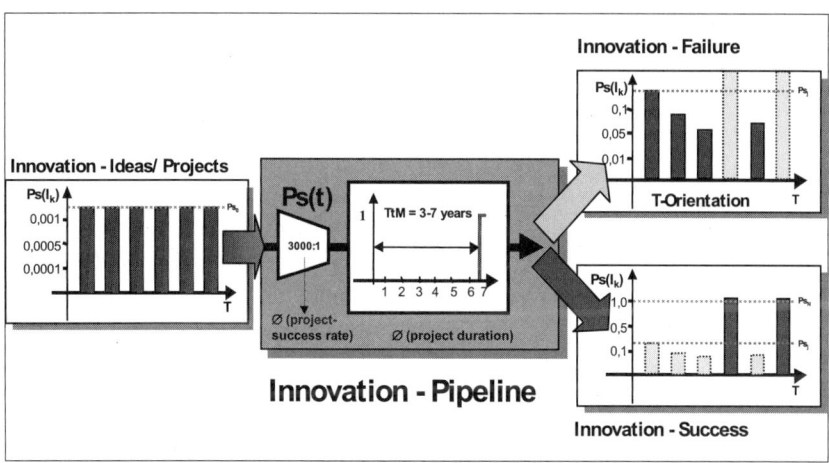

*Fig. 5: Basic control model for an innovation pipeline based on Lemma 3*

Assumed this most simple control model is correct, it renders an appropriate system description for some of the most basic control characteristics of inno-pipelines. So the next important question is:

*(Q 2)    How can one influence and control the respective output of an innovation pipeline?*

Just by looking on in Fig. 5 we immediately see that there are just three basic control strategies $S_c1 - S_c3$ to influence the output of an innovation pipeline as described and discussed below.

---

**Rule 1** - *the 3 most basic inno-pipe control strategies $S_c1$ to $S_c3$*

**$S_c1$)**    **Selecting the "right" input** innovation ideas, which is equivalent to modifying the acceptable "starting risk" $Ps_0$.

**$S_c2$)**    **Modifying the selection / success function $Ps(t_i)$** to make it more "selective" on the (supposed) innovation successes.

**$S_c3$)**    **Modifying the delay time** TtM of your filter.

---

### To $S_c3$ – Modifying the delay time:

It is most obvious that strategy $S_c3$ is **just giving you the results earlier not other nor any better ones.** So the output stays the same. It just makes your system (innovation pipe) run faster, but in general it forces you to do a major reorganization effort with a most doubtful result. Only do this if it is really necessary, for you will run into considerable problems. For a more detailed discussion on the kinds of problems one will most probably run into while pursuing such a strategy, please read [4].

### To $S_c2$ - Modifying the selection/ success function $Ps(t)$:

This strategy is the **most difficult but, at least in theory, the most promising one**. It just has a most severe disadvantage:

- Where do you draw the knowledge from to select the right innovation successes,

and if you can do so,

- why didn't you do so while constructing your innovation pipeline for the first time?

The underlying reason for this is,

**$S_c2a$)**    innovation-success (D 1) is an ex-post definition and

**$S_c2b$)**    one never will know the future in advance.

In spite of this problem this strategy remains a promising one, cause there are some ways to work around the problem of insufficient knowledge on the future (see [5] and especially chapters 3, 4).

**To $S_c1$ - Selecting the "right" input:**

This is the most widely used approach. It is the underlying concept of almost any portfolio selection and optimization approach to R&D. It just has two most simple problems:

**$S_c1a$)** It **does not work** due to argument $S_c2a$) and $S_c2b$) and even worse

**$S_c1b$)** It in general **produces fluctuations** because one now has constructed a fed-back delay system which oscillates almost by necessity as control theory predicts.

The only option to overcome this implicit fluctuation problem ($S_c1b$) is, just as control theory tells us, to **introduce some "predictor" for future innovation success into the feed-back system.** The best predictor we could imagine is an excellent chief engineer or an inventor-entrepreneur, just like T.A. Edison, G. Daimler or Bill Gates, with an excellent feeling for future market demands. The really important question with respect to that is,

**"Do you have such a person?",**            and if you do,
**"How do you recognize her/him to be the one?"**

Let us assume now that we do apply some reasonable control strategy to maintain our innovation pipeline IP in a good and stable condition. How can we ensure that our innovation pipeline is an economically reasonable endeavor, that it is at least able to pay for its own costs? Again an inspection of our control model (Fig. 5) helps. Obviously only innovation projects with Ps(Ip)=1 do earn money. All the other projects only do burn money. But, please do remember that, they are a necessary by-product of our filter or selection function Ps to find the few inno-successes! The above said does allows us to introduce the principle of the costs of information to optimize an innovation pipeline IP:

*Lemma 6 - the principle of minimal costs of information*

*(L 6)    The filter function Ps of an innovation pipeline IP does separate a set of innovation projects $\{Ip_1....Ip_n\}$ into two sets of projects:*

- *n-k innovation failures     IF $=\{Ip_1.....Ip_{n-k}\}$*
- *k innovation successes      IS $=\{Ip_{n-k+1}......Ip_n\}$*

*For any economically successful innovation pipeline the costs of information Ci*

**Ci = Costs (IF) < Profits (IS)     should be minimal!**

**Proof of Lemma 6 – the principle of minimal costs of information:**
There is no need to formally proof Lemma 6 cause, once one accepts our filter paradigm ((L 1), (L 2), (L 3)) and the corresponding control model (see Fig. 5), it is most obvious that only the inno-failures from set IF correspond to the economic losses of any inno-pipe IP. The inno-successes from set IS on the other side do correspond to the investments and their respective returns. Thus our **Lemma 6 is** nothing but **another form of the** most well known **golden rule of any economy** or economic endeavor:

- **"Try to maximize your profits, but at least do minimize your losses as good as you can!"**

The only minor precondition we have to make to even formally derive the principle of minimal costs of information is the assumption that one does not waste resources while pursuing the inno-successes down any inno-pipe IP (see below).                                                                **Q.E.D.**

This principle is the **basic optimization rule for any innovation pipeline**, cause given a certain innovation pipeline IP, its characteristic success function Ps(IP) determines the fixed number of the respective successes and failures. Because it is, by definition, assumed that each success, once you have finally found him, renders its maximum profit, you can only optimize the costs of the necessary trials/failures to identify these innovation successes.

The optimization of the respective profits of each innovation success is an integral part of normal product management and thus better should not be considered by any theory on innovation management.

**Remember:**
You cannot reduce the average number of innovation failures of a given innovation-pipeline IP. You just have to accept them. Doing this would result in an exchange of its characteristic filter function Ps(IP) which is equivalent to a complete redesign of your innovation pipeline into some new one. Now everything starts from scratch again and there is absolutely no reason to believe that your economical performance will be any better.

If carefully applied to the statistics of an individual innovation pipeline IP the principle of the **minimization of the costs of information is a most powerful tool** to design and **to optimize** complete **innovation pipelines** on a macroscopic, a mesoscopic and, with some additional help coming from the principles of knowledge management (see W.Klein in [5]), on a microscopic (project) level too. This will be part of the following chapters. We now have reached the end of the more formal description of our new innovation-process model. So let us summarize what we know about innovation so far on the following pages.

## 1.4    Intermediate summary 1 – the inno-gem[iv]

**S1a)**  Every **innovation project Ip is an investment project** too and thus it should pay just like T.A. Edison[v] already knew (see **(D 1)** and **(L 1)**)

**S1b)**  An innovation project Ip is a sequence of steps leading from some idea for a product/service to its respective coinciding technological (Ts) and market success (Ms):
**(D 2)**        $Ip(I_k) = \{Ip\ (t_0)........Ip(t_{N-1}), Ip(t_N)\}$
with        $Ip(t_N) = Ts(t_N)\ \&\ Ms(t_N)$

**S1c)**  The probability of an innovation success Ps is always the product of the respective technological (Ts) and the market success (Ms) probabilities as shown in

*Lemma 7 - the success probabilities Ps(IP):*

$$
\begin{array}{lllll}
(L\ 7) & Ps(Ip) = & Ps\ (Ts(Ip)) & *\ Ps(Ms(Ip)\ |\ Ts(Ip)) & (1) \\
 & = & Ps(Ts(Ip)\ |\ Ms(Ip)) & *\ Ps\ (Ms(Ip)) & (2) \\
 & = & Ps\ (Ts(Ip)) & *\ Ps(Ms(Ip)) & (3)
\end{array}
$$

(1) This describes the **technology-push mode** of innovation. It is proportional to the technological success probability times the market success probability for this technology.

(2) This describes the **demand-pull mode** of innovation. It is proportional to the market success probability times the technological success probability for this market.

(3) This describes **incremental innovations**. There one can assume that market and technology development are independent from one another:

**S1d)**  The probability of any innovation success decreases exponentially with "Time to Market" (TtM) or with the number of steps necessary to reach technological and market maturity:

$$(L\ 3)\ Ps(Ip) \sim exp(-\lambda * TtM(t_i))$$

Here the time $t_i$ is always measured backwards from completion of the respective innovation!

**S1e)**  **From a macroscopic (outside) view**, every innovation project Ip and its corresponding **innovation pipeline IP** can be described as a (technical) **filter process having an exponential filter function** (see S1d) and some characteristic delay or "filter time" (TtM)

---

[iv]    **inno-gem** for **inno**vation-process **ge**neral **m**odel
[v]    Thomas Alva Edison (1847-1931): "An invention that does not sell, I just do not want to make"

**S1f)** From a **microscopic (innovation project) view**, every **innovation project Ip** can be described as a "**shortest path" search-problem in its corresponding logical proof tree** (Herbrand Universe) for its proposition of coinciding technological (Ts) and market success (Ms).

$$(\mathbf{D\,2})\quad Ip(I_k) = \bigcup_{0}^{N} Ts_i(I_k)\ \&\ Ms_i(I_k)$$

**S1g)** **Macroscopic and microscopic view of the innovation process are complementary.** Delay or "filter time" translates into tree depth (number of steps) and the exponential filter function into the corresponding number of paths/ nodes in the respective proof tree.

**S1h)** In order to control an inno-pipeline most optimally, it is necessary to introduce a new, **meso-scopic view, the control layer.** This completes the inno-pipeline process model and allows **to appropriately describe and model management action and the respective response of an inno-pipeline**, being a statistical filter and a delay-time system at the same time!

**S1i)** The delay-time characteristic of any **inno-, R&D- or investment-pipeline** does make these systems **extremely sensitive to feed-back (learning) control strategies**. With the exemption of the introduction of some "intelligent predictor" any other control strategy applied does lead to harmful oscillations of the whole system.

**S1j)** The principle of the **minimization of the costs of information** (inno-failure costs) is the most general and the **most effective optimization strategy** for any inno-pipeline.

These 10 statements S1a) to S1j) fairly well describe the most basic properties of any innovation-process and thus may, only as a first order approximation naturally, be a sufficient answer for the first question:

**(Q1) "Which are the basic properties of an innovation process?"**

The results achievable with this inno-process model are even much better than one could expect at a first glance:

- With this complementary 3-level inno-process model/description - let's call it "m-cube model" ($m^3$-model) or the **"inno-gem"(vi)** – we now do have a complete, systematic and sound theoretical foundation to enter a most thrilling learning cycle (see fig.6) to further improve, extend and refine our knowledge on the innovation process.

---

vi    **inno-gem** for **inno**vation-process **gen**eral **model**;
    **$m^3$-model** for **m**acro-, **m**eso- & **m**icroscopic innovation-process **model**

- This model does render a lot of concepts and tools to make the knowledge rendered by management science operational and thus applicable for every day innovation- or R&D-management.

- This model allows for a new partnership between management science and every day innovation- or R&D-management in the different companies. This is possible, cause it is intrinsically testable and thus decidable. On top of this, it allows, once carefully applied, to most clearly distinguish between those aspects of innovation management one can generalize and the ones that are application (company, technology, market, etc.) specific.

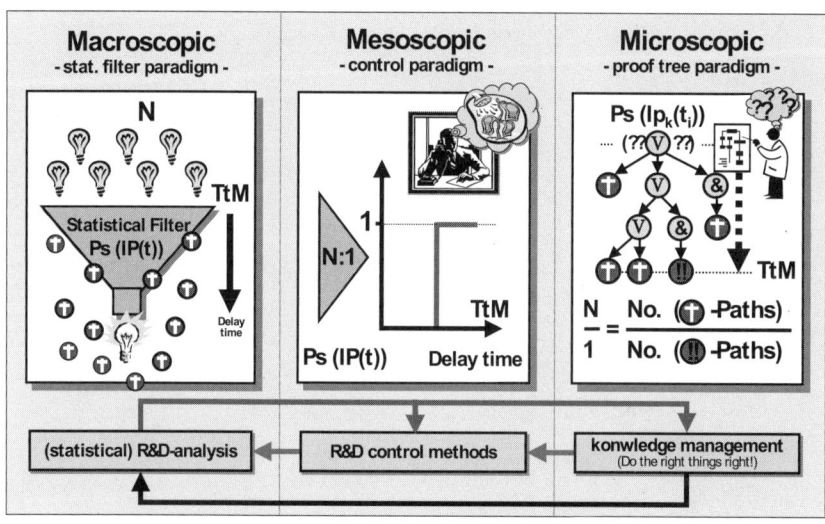

*Fig. 6: The innovation-process general model (inno-gem) and its learning cycle*

# 2. THE MACROSCOPIC VIEW ON REAL WORLD INNOVATION PIPELINES

As already mentioned in chapter 1, any inno-pipe IP could just be considered in general as an ordered set of N inno-projects IP={Ip(I$_1$), ... , Ip(I$_N$)} (see (D 3)). This generalization is by no means obvious, cause in reality we are discussing here highly structured and organized R&D systems, whether or not they are privately or state owned and operated. For the inno-gem, it is most important to know how this set of inno-projects IP={Ip(I$_1$), ... , Ip(I$_N$)} is constructed, structured and operated. Cause there are virtually infinite possibilities to construct such an inno-pipe, we will only discuss the two for practical purposes most important groups of inno-pipes. These are the (not gated) input-planning/selection and the gated/phased inno-pipes. We will deliberately leave the discussion and evaluation of other ones to more mathematically and (socio-) economically gifted authors.

## 2.1   From inno-projects to inno-pipes – how to construct IP(TtM (t))

A quite common way to operate an inno-pipe is sketched in Fig. 7. There you see how "normal", annual R&D-project planning/selection and control does produce an inno-pipe very much of the kind  described or assumed by the inno-gem (see (D 3), (D 4) and (L 3)). The reason behind this is the implicit "ordering" of inno-projects IP={Ip(I$_1$), ... , Ip(I$_N$)} by annually selecting and launching new "promising" ones. This leads to sets of projects in (almost) the same maturity state, which are more or less jointly migrating/aging down their (individual) maturity stages (milestones and/or quality gates) until they either become an inno-success or they are sorted out as an inno-failure. Thus their joint and inherently exponential success probability statistics (L 2) do, on an TtM-scale naturally, reproduce an equivalent TtM-exponential success probability statistic Ps(IP(t$_i$)) $\approx$ exp (-$\lambda$*TtM(IP(t$_i$))) for the whole set or the inno-pipe as a whole.

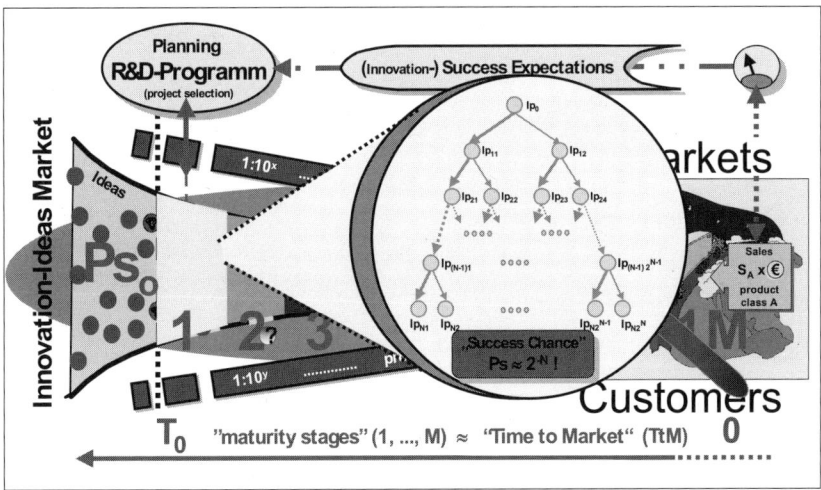

*Fig. 7: Sketch of a "R&D-input planning" inno-pipe ($\Leftrightarrow$ "select the right input strategy")*

Thus we can state the following lemma for any inno-pipe or R&D-system operated according to the "selecting the right input" strategy S$_c$1 (see also chapter 1.2):

*Lemma 8 - the exponential TtM-dependency of Ps(IP)*

*(L 8)*    ***Any inno-pipe operated according to the "selecting the right input strategy" does reproduce on a statistically determined TtM-scale the exponential time/TtM-dependency***

$$Ps(IP(t_i)) \approx \exp(-\lambda * TtM(IP(t_i)))$$

of its underlying set of individual inno-projects IP=$\{Ip_1, \dots, Ip_N\}$, provided the variances of the individual $\lambda_k$-values and the corresponding start-risks $Ps_0(Ip_k)$ of the set or of subsets of IP are not beyond the limits of statistical significance for the whole set IP.

---

**Proof of Lemma 8 – exponential TtM-dependency of Ps(IP) for input-planning inno-pipes:**
We will leave this proof to the interested reader. Referring to the above said, (L 8) is nothing but the logical consequence of a **time/TtM-respecting aggregation of the weighted exponential success probabilities of its underlying** and annually renewed **set of N inno-projects $Ip_k$ of the respective inno-pipe IP** under investigation.                                                    **Q.E.D.**

---

Naturally there is an infinite abundance of ways to aggregate different inno-projects $Ip_k$ to form an arbitrary inno-pipe IP. We will thus not go into more detail how such aggregations might be organized nor how the exact aggregation mathematics work in detail. What is really important to inno-management is solely the fact, that the respective inno-pipe does show an exponential success-chance or risk dependency with time better with "Time to Market" (TtM). This is true for an overwhelming majority of most of the practically relevant cases. The ones covered by Lemma 8 are, according to our experience the by far largest group. They are thus for real world inno-management the most important ones too.

There still remains another important group of inno-pipes. It is not quite as large or as important as the first one, but it is of special importance for good inno-management (see also chapter 3.4 inno-phase control). Thus we would like to investigate this group, the gated or phased inno-pipes, a bit more in detail.

## 2.2   The multiple-entry (gated or phased) inno-pipe

Looking a bit more in detail into gated inno-pipes one immediately sees, that they are nothing but a sequence of shortened (not gated) "input-planning inno-pipes" as already covered by Lemma 8 above. For the whole chain, naturally there could be some most radical changes in the inno-pipe success-function depending on the input-selection behavior or function being applied to the new or continued projects for the respective next phase.

Thus we have to state that for these (gated) inno-pipes in general Lemma 8 does not hold and that they could virtually show almost any time and/or maturity (TtM-) dependency of their respective inno-success probability depending solely on the input-selection algorithm for new/continued projects.

The really important question here is not whether or not one could construct a or any inno-pipe success-function **Ps (IP(t))**, but what are realistic or to at least some extend "reasonable" input-selection algorithms being or to be applied? Once we pose the question that way, we immediately do see too, that in general we will reproduce a sequence of more or less exponentially increasing success-functions for each sub-pipe or element of the chain of inno-subpipes. As we will point out in the subsequent chapter 2.4, the optimal choice for the respective "inno-speeds" ($\lambda$-values) of the elements (sub pipes) of the overall inno-pipe is the same constant value for all parts. Thus again Lemma 8 and the exponential time/maturity dependency of the inno-success function **Ps (IP(t))** holds. These thoughts do hold for the overwhelming majority of the in the real business world relevant cases of gated or phased inno-pipes.

## 2.3   General economic optimization rules for innovation pipelines

Following Schumpeters definition of innovation and own definition of an innovation pipe (D 3) it is must obvious that a necessary condition to run an inno pipe IP (e.g. a R&D-system) is the economical stability condition for its respective profits/losses PF(IP).

*Definition 7 - the economic stability condition*

$$(D\ 7) \quad PF\ (IP\,(\Delta t)) = \sum_{1}^{N} EVA\ (Ip_{i}\,(\Delta t)) \geq 0$$

*with respect to some arbitrary time period $\Delta t$.*

Starting with the economic stability condition (D 7), applying the statistical filter paradigm ((L 1), (L 2), (L 3)) and the cost of information principle (L 6), one can derive the following inequation to describe the fundamental economic optimization threshold condition for any innovation pipeline IP:

*Lemma 9 - the fundamental inequation for the economic success of an innovation pipeline*

$$(L\ 9) \quad \overline{Ir}(IP,\Delta t) > \frac{\overline{cos\,ts}(Ip_{k} \in IF(IP,\Delta t))}{EVA_{S}(Ip_{k} \in IS(IP,\Delta t))} = \frac{\overline{C_{F}}(IF(IP,\Delta t))}{EVA_{S}(IS(IP,\Delta t))}$$

With   $\overline{Ir}(IP,\Delta t) = \dfrac{Number\_of\_elements(IS(IP,\Delta t))}{Number\_of\_elements(IF(IP,\Delta t))}$

*average innovation-rate of the inno-pipe IP*

and   $\overline{C_{F}} = \overline{cos\,ts}(IF(IP,\Delta t))$

*average cost of an innovationfailure   $Ip_{k} \in IF(IP,\Delta t)$*

and   $\overline{EVA_{S}} = \overline{EVA_{S}}(IS(IP,\Delta t))$

= *average Economic Value Added (EVA) of an innovation success   $Ip_{j} \in IS(IP,\Delta t)$ with respect to an arbitrary time period $\Delta t$*

> **Proof of Lemma 9 – fundamental inequation for the economic success of an innovation pipeline:**
> We will leave this proof to the interested reader, cause (L 9) is nothing but a statistical and mathematical transformation of the economic stability condition (D 7) applying the filter-paradigm ((L 1), (L 2) and (L 3)) and the corresponding control model (see proof of Lemma 6).        **Q.E.D.**

Any economically stable innovation pipeline must fulfill this fundamental inequation, cause always remember, only the innovation successes $Ip_j$ from set IS and also only their discounted net profits, with all costs deducted, can/must pay for the whole investment-, R&D- or innovation-pipeline. The rest, all the projects $Ip_k$ from set IF just burn money.

The maximum output one can get from the results of the projects $Ip_k$ from the inno-failures set IF is technical orientation, information which helps to answer the question:

### "How we can do better next time?"

In almost any real inno-pipe or R&D-system this information is mostly discarded. This is done although this information would be most beneficial, once properly stored, coded and used. Personally we are very much inclined to claim that the proper use of this information represents the biggest ratio-potential in any innovation process at all. It is just not very easy to tap on it. That is the key problem, as we will see in more detail, in the discussion of the microscopic project-view in chapter 4 (see also [5] for additional information). We will not proof Lemma 9, for it is a most simple calculation to demonstrate, that this lemma is just a logical consequence of the stability condition (D 7). Now let us discuss the fundamental inequation (L 9) for an inno-pipe a bit more in detail:

a)   The three figures **innovation-rate Ir, failure-costs $C_F$** and the **net profits** of the **innovation-successes $EVA_S$** are the **decisive set of performance figures** for any innovation-pipe, R&D- or even an investment-system or -process.

b)   **Ir, $C_F$ and $EVA_S$ are interdependent.** Trying to increase Ir undoubtedly increases in general $C_F$ and decreases $EVA_S$, cause you now have to go for the not so thrilling chances too.

c)   The **time constants** of these performance figures (**Ir, $C_F$, $EVA_S$**) are quite **different**:

- $C_F$ has the **shortest one**, you know the costs of the stopped projects immediately.

- **EVA$_S$** is only known (exactly) at the end of the respective product life cycles, thus it has the **longest time constant.**
- **Ir is in between.** You need not wait until the EVA$_S$ of the respective successes are known exactly, but you must be sure that the respective innovation projects do produce an EVA.

A short inspection of the fundamental inequation (L 9) of the innovation business tells us that there are only **3 basic strategies to optimize the economic performance of any inno-pipe** or -system:

| Rule 2 - *the 3 most basic inno-pipe economic optimization strategies $S_e1$ to $S_e3$* |
|---|
| $S_e1$) **Maximize the innovation rate Ir(IP)** |
| $S_e2$) **Minimize the failure costs C$_F$( IP) = C$_F$( IF)**  (see (L 6)) |
| $S_e3$) **Maximize the net profits of the innovation successes EVA$_S$(IP) = EVA$_S$(IS)** |

**To $S_e1$) - maximize the innovation rate Ir (IP):**
This is a typical economic success strategy for young dynamic markets and (product-) technologies. It closely corresponds to M. Porters "technology leader" strategy, where the profitability stems from temporal monopoly profits due to (technologically) produced USP´s.

- If you cannot sustainably differentiate yourself from the competition with your products do not try this strategy.
- If you do not master the necessary key- and, in case, the respective brake-through technologies too, do not try either.

The ability to maintain a sustainable technology and/or product differentiation capability is the decisive precondition to go for such a "Premium-Strategy".

**To $S_e2$) - minimize the failure-costs C$_F$ ( IP) of the set of innovation-failures IF:**
This is not a classical cost cutting strategy. The essence of this strategy is to stop "hopeless" innovation projects Ip$_k$ as early as possible and to reinvest the saved resources into new, more thrilling ideas. Doing this prevents the resources from being wasted and thus your relative search costs per innovation success C$_S$(IP) are statistically minimized. This overcomes the most severe cloven hoof of standard cost cutting strategies in the innovation-, in the R&D- and in the investment-business:

- They in general are much more effective in damaging and in preventing an innovation success from taking place, than in reducing the costs of the investments necessary to survive economically.

Used properly (see comments to (L 6)), as we will describe in more detail in the chapters 3 and 4, this strategy is the most powerful approach to reduce the necessary investment (costs) to produce 1Euro of innovation-success net profit ($EVA_S$). This is due to 2 main reasons:

**Se2a)  Minimizing the failure-costs $C_F$ does hardly affect the innovation rate Ir( IP)** and even if, it rather has the tendency to increase it, due to the reinvested resources.

**Se2b)  Minimizing failure-costs $C_F$ does not affect $EVA_S$,** cause the investments in the thrilling ideas are not affected at all. Additionally there is no reason to believe, that the chances to identify innovation-successes are diminished, once your "show-stopper criteria" are carefully selected and monitored (see chapters 3 and 4).

**To $S_e3$) - maximize the net profits $EVA_S$ (IS) of the set of innovation-successes IS:**
This strategy can be executed in 2 most different ways:

**$S_e3a$)  Maximize the (discounted) net life cycles profits ($EVA_S$)** of your product-innovation successes IS.

**$S_e3b$)  Maximize the "expectancy value" of the $EVA_S$** of your product-/innovation-successes IS and sell them off as early and as good as you can.

**To $S_e3a$) - maximize the (discounted) net life cycle profits $EVA_S$(IS):**
This is nothing but the normal job of any product management. There are quite some techniques on the market and there is literature on that topic in abundance. Thus there is no reason to reinvent the wheel. We just have to assure that product management is doing its job right. This is really not a topic of nor a question specific to innovation management.

**To $S_e3b$) - maximize the "expectancy values" for the $EVA_S$(IS) and sell them off:**
This is a most typical strategy for the VC-industry. It works quite well, but it has the problem of cyclical market, better cyclical investor expectations or "investment fashions".

This is not a strategy accessible for a large firm or company, which wants to make a sustainable long term business other than investment/financing. Even for investment firms this strategy is a quite risky one, cause supposed you go for the big profits, 10 or even some 100 times your investments, how do you guarantee that the next "big deal" is there in time before you yourself go bankrupt? Your only chance is to have enough "promising horses" in the race and this requires very substantial financial assets again. To summarize our discussion of the fundamental inequation

$$(L\ 9) \qquad \overline{Ir}(IP, \Delta t) > \frac{\overline{C_F}(IP, \Delta t)}{\overline{EVA_S}(IP, \Delta t)}$$

we may state the following **"two golden rules"** or strategies **for innovation management:**

---

*Rule 3 - two „golden" rules $S_I1$ and $S_I2$ for good innovation management*

$S_I1$) **Keep your innovation rate Ir( IP) as high as you can afford, but cut down on your necessary search costs $C_F$( IP) as much as you can!**

$S_I2$) **Keep the money saved by stopping "hopeless" projects in your innovation pipeline and reinvest it such, that you maintain a stable flow of innovation projects in your filter!**

---

It is a little more difficult to monitor whether or not you are still on track with this strategy than writing it down, cause the performance figures Ir, $C_F$ and EVA$_S$ are nasty to determine. But it is for sure not impossible and the inertia of your R&D- or your inno-pipe does help you quite a bit. This inertia, once you have set up an appropriate monitoring system, does enable you to do good estimates for the future development of these performance figures by analyzing their respective histories:

- For **$C_F$(IP, $\Delta t$)** you get values updates once you stop the respective "hopeless" projects $Ip_k$. This is the performance figure **almost "à jour"**.

- For **Ir(IP, $\Delta t$)** this is a bit more difficult, cause before you call some innovation project $Ip_j$ a success ($Ip_j \in IS(IP, \Delta t)$), you should at least wait until product marketing successfully starts. Thus on average the values for Ir(IP, $\Delta t$) are always about **50% of** an average development time **TtM$_D$ behind.** You just have to extrapolate for this time to compute the inequation (L 9).

- **EVA$_S$ (IP, $\Delta t$)** is only known at the end of the respective product life cycle $T_{LP}$. Again on average you have to extrapolate the respective values for EVA$_S$ for about **50% of** your average **product life cycle time $t_{pc}$.**

From the three basic strategies ($S_c1$ to $S_c3$) you see how statistics does help to run and to optimize an innovation-pipeline. It is good to know your statistics and especially their respective development trends, but e.g. if your statistical basis is too small, a look to your industry and to your competitors will help. Thus benchmarking is definitively not obsolete, it is even more important and beneficial, once you use the fundamental inequation to compare the respective results, figures and performances achieved.

## 2.4   How to design optimal innovation pipelines?

In Chapter 2.1 we saw how our filter and control model together with the principle of "minimized costs of information" did lead us to the fundamental inequation. This is the global economic success condition to run an innovation pipeline. We still do not know an answer to the especially for R&D-dependent companies most important question:

> *(Q 3)   How is an optimal innovation pipeline structured and organized?*

We assume, any careful reader of chapter 1 will almost guess the answer to this question. It is really most simple, once one really accepts the validity of the filter paradigm. Just think of a technical filter process, which is a most useful analogy to an innovation-pipeline. For any technical filter process, there are 2 most crucial optimality conditions to be fulfilled. Exactly these two conditions hold for innovation pipes as well:

*Lemma 10 - the matching condition for optimal inno-pipe capacities*

$$(L\ 10)\quad \sum_{j} C_{out}^{j}\ (IPstage\ (k)) \le \sum_{i} C_{in}^{i}\ (IPstage\ (k+1))$$

*The maximum output capacity $C_{out}$ of an arbitrary stage $k$ of an optimally designed inno-pipe IP must not exceed the input capacity $C_{in}$ of the respective following stage $(k+1)$!*

*Lemma 11 - the matching condition for optimal inno-pipe progress*

$$(L\ 11)\quad Ps\ (t_k)/Ps(t_k+\Delta t_k) = C_P \approx ln\ (N_k/N_{k+\Delta})\ /\ \Delta t_k$$

*with $N_{k+\Delta}$ = number_of_projects in the inno-pipe IP at $TtM = t_k + \Delta t_k$*

This progress or speed match is the direct correspondence to the constant flow of material condition through a technical filter. This is especially important, for gases, because they are compressible. The same condition holds for inno-projects too. An optimal flow of ideas (corresponding to a constant pressure gradient in a technical filter) is only guaranteed, if there is a constant speed of "complexity reduction" or of augmentation of your respective chance for an inno-success along the respective TtM-scale of your inno-pipeline.

**Proof of Lemma 10 – capacity match:**

The proof of Lemma 10 is really most simple, almost self-evident: Suppose $C_{out}$(stage k) > $C_{in}$(stage (k+1)). Then you most certainly are not able to pursue some m inno-projects IF = $\{Ip_{k1}, \ldots, Ip_{km}\}$ due to capacity reasons. As a consequence you will loose some n inno-successes IS = $\{Ip_{k1}, \ldots, Ip_{kn}\}$ and their respective $EVA_S$ (IS). But, quite in contrast to the other "normal" continuing inno-projects $Ip_k$, you already have paid (invested) for (in) their evaluation until the mismatching stage k. Thus you just maximized $C_F$(IF) and minimized Ir(IP) without changing the respective $EVA_S$(IP). This is clearly not economical (see also (L 9)) and thus suboptimal.     **Q.E.D.**

**Proof of Lemma 11 – speed match:**

For this proof, there is a formal, more complicated but more scientific way using Ps(t) $\approx$ exp (-$\lambda$*t) and a phenomenological, more simple way. We will take the latter one and leave the other to the interested reader:

Suppose there is some $\Delta t_k$, where you do achieve a much bigger progress in your success function Ps. Now it is an optimal choice just to continue with that speed of "complexity reduction" across all your inno-projects $Ip_k$ towards your goal, the end of your inno-pipe and finally the completion of your innovations IS(IP).

The consequence of this "optimal choice" is nothing but a different value for the speed parameter $C_P$. After having "optimized" the whole inno-pipe IP according to our new "speed-strategy", we finish with just another constant value for our speed-parameter $C_P$. This again is nothing but another constant speed $C_P$ for our now newly "optimized" innovation-pipeline IP.     **Q.E.D.**

It is most important for an optimal inno-pipe design to respect these 2 optimality conditions for speed (L 11) and for capacity (L 10). There are still quite a few companies around, who think they need not pay respect to them. They do and they will pay real money for it.

In vehicle development Toyota and Honda are very often taken as benchmarks [6]. The capacity distribution for their research, pre- and serial-development departments did and does follow an almost exponential curve. This does nicely correspond at least to some extend to our theoretically derived optimality conditions. Perhaps some theories really do describe, what we would like to call reality?

Now let us continue with our quest for the design rules for optimal inno-pipes. Again our technical analogy, the filter paradigm, shows us just by inspection, that there are only 3 basic strategies ($S_o1 - S_o3$) to structure and organize an inno-pipe (see Fig. 8):

| Rule 4 - *the 3 most basic inno-pipe organization principles $S_o1$ to $S_o3$* | | |
|---|---|---|
| **$S_o1$) Cascading** | - | divides a given inno-pipe into some k fractions of itself. |
| **$S_o2$) Buy-in** | - | adds at some stage k the input from some outside inno-pipe IP |
| **$S_o3$) Paralleling** | - | just operates several inno-pipes in parallel with no interaction of their respective pipes, inputs or outputs. |

It is obvious, that our 2 basic optimality conditions ((L 10) capacity- and (L 11) speed-match condition) should be respected once executing each of these 3 basic inno-pipe design rules. This has rather drastic consequences, as shown in Fig. 8, for the design parameters of the respective stages for each basic design. These consequences, together with the respective basic inno-pipe designs, will be discussed in the next paragraphs.

### To $S_o3$) – paralleling an innovation-pipeline IP:

It is most obvious that this is just an aggregation of any combination of $S_o1$- to $S_o2$-designs for different products, markets, technologies or firms. So there is no interaction between those pipes and the respective in- or outputs just add up. Conflicting needs for shared resources are neglected here. To resolve them, normal operations research does offer more than elaborate tools, and again, we should not reinvent the wheel! Thus optimizing each inno-pipe separately while respecting the A/B-product optimization rules from classical operations research is the right way to run such an innovation-system or -process.

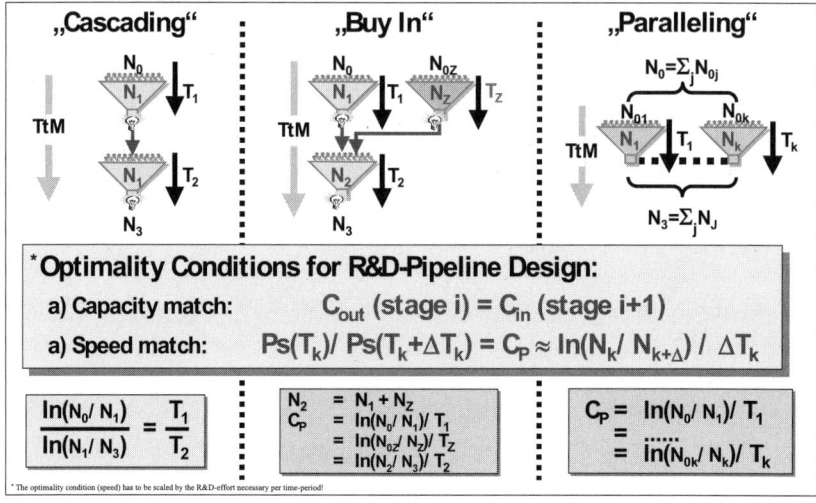

*Fig. 8: The 3 basic innovation-pipeline designs and their respective optimality conditions*

## To $S_o2$) – buy-in:

This is the predominant inno-pipe design for assembly industries like e.g. the car and the aerospace industry. There you cannot do everything yourself, you just have to integrate the (component-) innovations of other industries into your product innovations. I&C-technologies, electronics and mechatronics are the major buy-in items for these, relatively old industries. Their products instead, are quite young and they are still developing most dynamically. So what are the basic design rules for this case (see Fig. 8):

$S_o2a)$    The capacities at each stage k just add up. So, most important, the input-capacity of stage k must fit to the sum of the internal and the external output-capacities of the respective previous stages (k-1).

$S_o2b)$    The maturity level, especially of the bought-in items ($C_{ext}$), must match. Once you have a quality gate system in place, the respective requirements must hold for external items too. So you just have to integrate your suppliers into your maturity/quality gate system as closely as your own previous stages of your inno-pipe!
This is a most management intensive task, cause your supplier will remain an outsider. Thus you will have only most limited possibilities to influence his behavior and, even less, his future actions and plans too!

**To $S_o1$) – cascading:**

Cascading is the basic design rule to get a hand on the controls of an inno-pipe being a statistical filter process. Conceptually this strategy corresponds to the introduction of stages in (technical) filter processes like e.g. refineries, distilleries or even a diamond mine.

The analogy to a technical filter process does really hold quite a bit. Thus, it might be a good idea for MBA´s doing innovation management to learn a little more about these processes and their respective design rules and optimality conditions.

Just by looking at technical filter processes we can answer the question – "what are the elementary design rules for cascaded innovation-pipelines?" – much better than by consulting traditional innovation management textbooks.

## 2.5 Design and optimization rules for cascaded innovation-pipelines IP

*Rule 5 - the 6 most basic design and optimization strategies $S_d1$ to $S_d6$ for cascaded inno-pipes*

$S_d1$)  **Never ever break the innovation chain:**

This is the most obvious and the most violated design rule too. If you break the chain, at least in a technical filter process, you don't get results (products) at all. For inno-pipes, at least their respective performance (the output/input ratio) is usually severely diminished, if you do not respect this design- ,operation- and control-rule for innovation management.

We cannot imagine the reasons, why some R&D-managers still believe that they can cut out parts of their inno-pipe, e.g. the predevelopment department, without severely damaging and/or altering the whole pipe with most doubtful overall performance results, at least in the long run?

$S_d2$)  **The weakest part of an innovation pipe limits its overall performance:**

This rule is most obvious and accepted for technical filter processes. For an inno-process this is the less dramatic version of "breaking the chain". The consequences are quite similar (see the proofs of (L 10) and (L 11)). Again you do have to pay real money for not respecting this design rule!

$S_d3$)  **Once the chaining-conditions are respected one can optimize every part of an innovation-pipe individually:**

This is a most useful property of any innovation-processes. The basic idea behind it is a most simple one.

Once you know what the next step in your inno-chain really wants/needs, you exactly know the innovation-market for the stage your in. Once you have knowledge of the output of the previous stages of your inno-pipe, you know your input.

So you know everything you need to optimize any stage of your inno-pipe individually. You now just apply the fundamental inequation (L 9), the capacity- (L 10) and the speed-matching (L 11) criteria. Having done that, you are almost finished with the optimization rules for the individual stage you are designing and managing. By the way, quality gate processes do apply and exploit this intrinsic property of any inno-pipe very much to the benefit of their designers.

**$S_d4$) Organize an innovation pipe according to its respective TtM- or inno-maturity levels:**

From the filter-paradigm ((L 1), (L 2) and (L 3)) we know that Time to Market (TtM) or the inno-maturity level achieved (see proof step3 for (L 2) and (L 3)) are the decisive parameters for the respective success-function Ps(t) of every inno-pipe.

Once you want to'optimize your inno-pipe, it is thus a good idea to organize it and its respective processes along these parameters. So the information generated inside the pipe will help you to organize and control it accordingly. We will further describe this approach in more detail in the following paragraphs.

**$S_d5$) The capacity distribution along its TtM- or inno-maturity-levels determines the efficiency potentials of the respective inno-pipe:**

This design rule is a most important and a most helpful consequence of the principle of "minimal costs of information" (L 6) for any inno-pipe. We will derive and line out some example capacity distributions and their respective costs/benefits using this and the previous rule ($S_o1$) in the following paragraph.

**$S_d6$) Organize and control the different filter-, technology- or product-stages according to their respective maturity levels and to their respective average throughput times:**

Now we finally reached the last optimization possibility/rule for the design of inno-pipes. Applying this design and control rule properly, is top of the notch innovation management. It is the ultimate optimization potential once you exhausted the possibilities of good innovation management, which is nothing but exploiting the respective application potentials of the rules $S_d1$ to $S_d5$ properly.

As an example application of the inno-process model (inno-gem) described in chapter 1, we will now apply this model and its respective lemmas to the design and optimization rules $S_d4$, $S_d5$ and $S_d6$.

### To $S_d4$ and $S_d5$ - optimal capacity designs Ca(IP) for a (cascaded) inno-pipes IP:

Following the filter paradigm (L 1) to (L 3) and the control model (see Fig. 5) it is possible to compute the costs of information according to Lemma 6 for any inno-pipe IP. This is possible, cause the pipe just produces 2 kinds of information, the inno-success info for the members of set IS(IP) and the inno-failure info for the members of set IF(IP). Probability calculus does render us the most helpful relation:

*Definition 8  -  the error probability*

$$(D\ 8)\quad P(IF)\ =\ (1-P(IS))\ =\ (1-Ps(t))$$

Together with an assumed capacity distribution $Ca_{IP}(t)$ along the TtM- or the inno-maturity scale of the respective inno-pipe, this relation does allow us to compute the costs of information this inno-pipe will produce on average. Here the assumption is made that the costs per inno-projects on average are more or less equivalent to the average capacity deployment inside the respective inno-pipe. According to our personal experiences, this assumption is a most reasonable one for most real inno-pipes. With these assumptions we may calculate the **cost of information Ci(IP)** produced inside an arbitrary inno-pipe IP as follows:

*Definition 9  -  the costs of information*

$$(D\ 9)\qquad Ci(IP) = \int_{0}^{TtM_{IP}} Ca_{IP}(t) * (1 - Ps(IP)) dt$$

$$= \int_{0}^{TtM_{IP}} Ca_{IP}(t) * (1 - e^{-\lambda t}) dt = C_{F}(IP)$$

With a corresponding set of definitions we may calculate as the respective **investments** we have to make **to generate** a corresponding set of **innovation-successes Cs(IP):**

*Definition 10  -  the costs of the innovation successes*

$$(D\ 10)\qquad Cs(IP) = \int_{0}^{TtM_{IP}} Ca_{IP}(t) * Ps(IP)) dt = \int_{0}^{TtM_{IP}} Ca_{IP}(t) * e^{-\lambda t} dt$$

With these definitions for the success-investments or -cost Cs(IP) and the failure- or information-costs Ci(IP) we may calculate the **inherent maximum efficiency $E_{IP}$(IP)** of different capacity designs $Ca_{IP}(t)$ for a given inno-pipe IP:

*Definition 11  - the maximum efficiency for a given inno-pipe*

$$(D\ 11)\qquad E_{IP}(IP) = \frac{Cs(IP)}{Cs(IP) + Ci(IP)} = \frac{Cs(IP, Ca_{IP}(t))}{Cs(IP, Ca_{IP}(t)) + Ci(IP, Ca_{IP}(t))}$$

Using the start-risk parameter $\alpha$(IP), we set the exponential capacity strategy parameter to $\lambda$(IP), which is by the way an optimal choice. Doing that, we get the following definitions (D 12) to formalize and calculate the effects of different inno-pipe capacity deployment strategies on their respective maximum efficiencies $E_{IP}$:

*Definition 12 - inno-pipe capacity deployment strategies*

*(D 12)* $Ps_0(\alpha)$ = $Ps(TtM_{IP}, IP) = exp(-\alpha)$ *is the start-risk parameter with*
$\quad\quad\lambda(IP)$ = $\alpha/TtM_{IP} \approx \alpha/t_{pc}$ *and the capacity strategies*
$\quad\quad Ca_{IP}(t)$ = $k_0$ *constant R&D-capacity or,*
$\quad\quad Ca_{IP}(t)$ = $k_0*(TtM_{IP}-t),$ *linear R&D-capacity increase or,*
$\quad\quad Ca_{IP}(t)$ = $k_0*exp(-a*(TtM_{IP}-t))$ *exponential R&D-capacity increase.*

We can now compute the following table (1) for the respective inherent inno-pipe efficiencies $E_{IP}$ for these 3 most common capacity deployment strategies $Ca_{IP}(t)$. Additionally Fig. 9 shows a plot of the respective (maximum) inno-pipe efficiencies $E_{IP}$ according to the "cost of information principle" (L 6).

| Start-chance $\alpha$: | $\alpha = 2$ (Ps$_0$ = 13,5%) | $\alpha = 3$ (Ps$_0$ = 5%) | $\alpha = 5$ (Ps$_0$ = 0,7%) |
|---|---|---|---|
| $Ca_{IP}(t) = k_0$ | $E_{IP} \approx 43,2\%$ | $E_{IP} \approx 31,7\%$ | $E_{IP} \approx 19,9\%$ |
| $Ca_{IP}(t) = k_0*(t_{pc}-t)$ | $E_{IP} \approx 56,8\%$ | $E_{IP} \approx 45,6\%)$ | $E_{IP} \approx 32,1\%$ |
| $Ca_{IP}(t) = k_0*e^{(-\lambda*(t_{pc}-t))}$ | $E_{IP} \approx 56,8\%$ | $E_{IP} \approx 52,5\%$ | $E_{IP} \approx 50,3\%$ |

*Table 1: Inno-pipe efficiency $E_{IP}$ as a function of the capacity deployment strategy $Ca_{IP}(t)$*

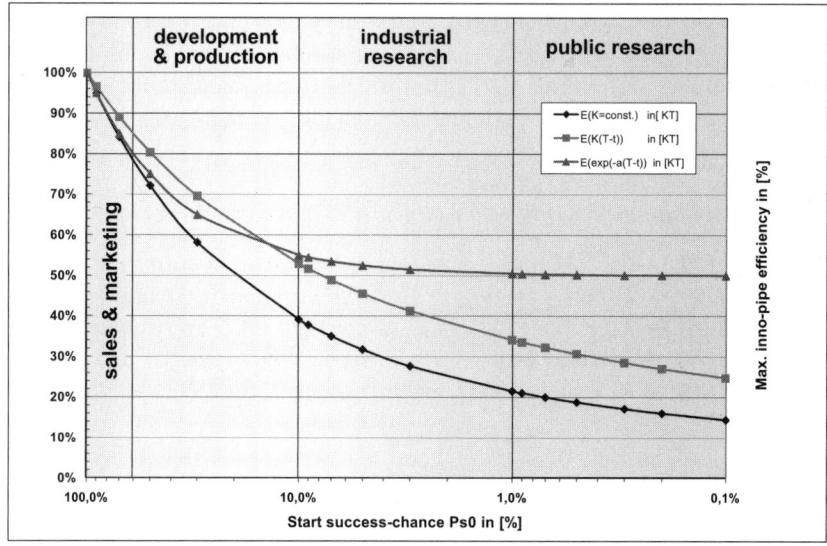

*Fig. 9: Plot of the inno-pipe efficiency $E_{IP}$ as function of start-risk $Ps_0$ and capacity strategy $Ca_{IP}$*

Fig. 9 most impressively demonstrates in numbers the potential efficiency benefits of an exponential R&D-capacity deployment along the innovation chain. This is especially important in the public and in the industrial research sector, where the biggest part of the "complexity reduction" or enhancement of the probability to become an innovation (see success-function Ps(t) in Fig. 3), has to be performed.

The following stages (industrial development, marketing, sales and after-sales) only have to increase the probability in general from some 10-20% to a full 100%, which is a factor of 5-10. Research, in contrast, usually has to deal with factors from 10 to over 100! Again, we cannot overemphasize the enormous efficiency and, as a consequence, the efficacy benefits too, of this R&D-capacity deployment strategy. Having Fig. 9 in mind, it is not astonishing at all, why Toyota and Honda are organizing their inno-pipes the way they do! By the way, we are most certain that a classical chief engineer, like T.A.Edison or G.Daimler, has done and would have done so too.

## 2.6  Design rules for cascaded inno-pipes - the pharma example

**To $S_d4$ - organize your inno-pipe according to TtM-levels (the pharma example):**

Very much in contrast to the automotive or the aerospace industry, you have in the pharma industry an almost one to one relationship between innovation investment (into a drug) and its respective profits or inno-success. So this industry is most suited to test the validity and the applicability of our innovation-process model and the corresponding theory.

The pharma process is a most simple, but a very long one. The **product development** of a new drug usually takes about **15 years** and has about a **1 out of 20 success** chance, which is not too good, compared to other industries. So you have to deal with a considerable risk and also with most remarkable **investments** necessary, about **3,6 Mio $ per drug on average**, before you obtain a new one. This situation is illustrated in Fig. 10 using our control model adapted to the (average) pharma industry drug development process.

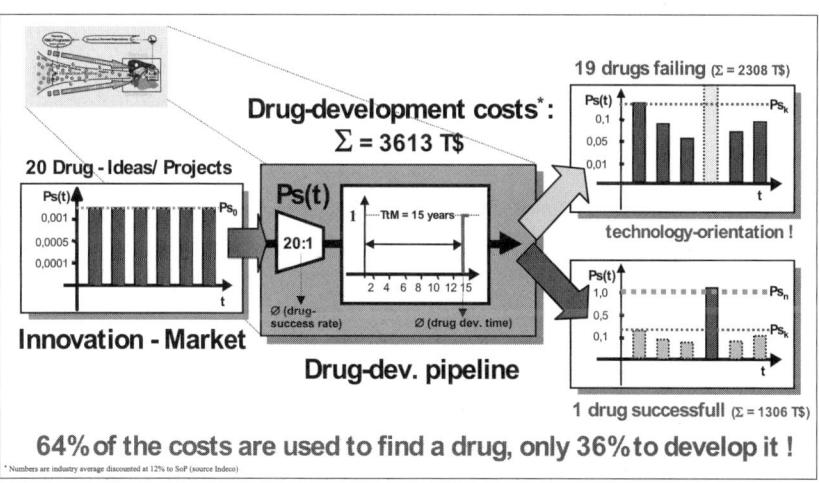

*Fig. 10 :Control model of the industry average pharma filter process (inno-pipe)*

Compared to our calculated maximum achievable inno-pipe efficiencies, the average **pharma industry development process does not perform too well** at all. It shows quite some room for improvement, cause for an average start-success chance of $Ps_0=5\%$, as in the pharma industry, we would expect some $E_{IP}=50\%$ inno-pipe efficiency, not a **rather meager $E_{IP}=36\%$**. Remember these are industry average figures. Thus, there are obviously quite a few companies performing even much worse!

---

*(Q 4)*    ***What are the deficiencies of the pharma R&D-process that it performs the way it does?***

---

To answer this question, we have to look a little closer to the drug development process, sketched in Fig. 11, and compare it to our inno-process model to determine, where things are going wrong. Now, what does Fig. 11 tell us about the pharma filter process:

**1.)** The pharma process is a **phased and gated** development process, with a TtM of about 15 years covering 6 distinct development phases. That is OK according to our theory and our model.

**2.)** The pharma development **phases are not equidistant**. This is not a problem in itself, but it does make management and control of the whole process (much) more difficult!

**3.)** The pharma process most obviously does severely **violate our speed-match** criteria and, most probably, the capacity match criteria too! This is by no means a smooth filter process. It rather resembles a **series of notch filters** (see red curve in Fig. 11), just like a comb. This is a **not at all suitable** solution for a statistical filter problem. This is common sense, at least for almost any control engineer. Thus, as already mentioned before (see comments to the proofs of (L 10) and of (L 11)), you have to pay substantial amounts of (real) money for not following these rules!

**4.)** Starting with year 10 (see blue curve in Fig. 11), corresponding to the **end of** (clinical) **phase 2, drug development is becoming really expensive,** more than 10 times the cost per year than before! Despite that, there is still, on average, a considerable risk, cost and uncertainty left in the process at that stage or phase.

**5.)** The **pre-clinical phase** is by far the major filter stage and thus the **most costly one**, where (potential) inno-failures are sorted out by the pharma process.

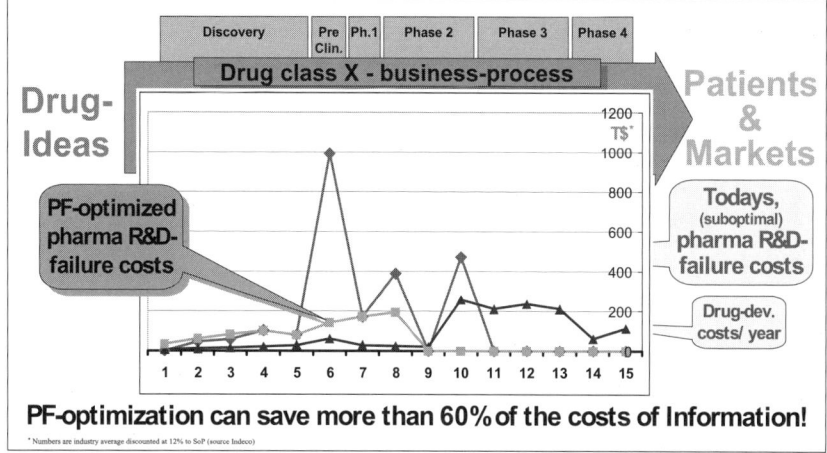

Fig. 11: Plot of the pharma inno-pipe (industry average) and its respective efficiency potentials

These sketched properties (1 to 5) of the drug development process do lead us to following question:

> **(Q 5)** **How can we improve the pharma (drug development) filter process?**

Again the main answer is a most simple one and it is already outlined in our **inno-gem** (see intermediate summary 1). The theoretically derived speed- and capacity-match criteria do tell us that only smooth, exponential filter designs are optimal. Anything else is nothing but expensive. So, why not replace the main filter stages "pre-clinical" (year 6), start "phase 2" (year 8) and end "phase 2" (year 10) by the most simple (darwinistic) filter approach one can imagine:

- **Until year 10, when pharma projects are still cheap, every year only a fraction (e.g. 70%) of the drug projects do survive their respective (annual) maturity tests!**

- **From year 10 on, when drug projects become real expensive, almost every drug project (e.g. >95%) does/must survive its necessary maturity test!**

Once we recalculate (using the inno-gem) the pharma process under these assumptions we see, that

- **more than 60% of the (necessary) costs of information $Ci(IP_{pharma})$ could be saved!**

- **The overall inno-pipe efficiency $E_{IP}$ could be increased by more than 50%!**

These (potential) savings are not at all taking into account further potential savings due to the application of other aspects (e.g. knowledge management, improved control, etc.) of the inno-gem. Thus there is quite some justification, that we may expect even a lot more room for improvement in the drug development process, than the calculated -60% from the costs to find the "right drug" $Ci(IP_{pharma})$.

Having these numbers in mind, considering the tremendous future market potentials of the "life sciences" in general and knowing the enormous impact of the (drug) development costs on the balance sheets of "big pharma" one might be tempted to promise this industry a rather bright future, once it really makes use of all these still unrevealed efficiency potentials in their innovation-processes!

## 2.7 Logical paralleling - the ultimate optimization method

**To $S_d6$** - organize and control the filter- (technology-, or product-) stages of an inno-pipe according to the respective maturity levels and to the average throughput times:

Let us now suppose we have organized our inno-pipe according to the inno-gem. Thus we have

a) created a steady and **continuous inno-pipe** without ruptures or weak-links.

b) **organized our inno-pipe along the TtM- or maturity scale** relevant for our product/market and we have selected a start-risk $Ps_0$ or a corresponding innovation height we are aiming at.

c) implemented a **complete and consistent quality gate scheme** to always be able to compute and to monitor the development of the success-function $Ps(t)$ of our inno-pipe.

d) set up **optimization schemes for each of the stages** of our inno-pipe.

e) always respected and incorporated the **speed- and capacity-match conditions** into the design of our inno-pipe, e.g. using an (optimal) **exponential capacity deployment strategy** (see Fig. 9).

Having done and achieved all that, we may ask ourselves now,

- **are we really done with the inno-pipe optimization possibilities doing steps a) to e)?**

The answer to that question is

- **yes, provided we do have a one product, one technology and/or one market problem only!**

These kinds of inno-problems (one product/market) are more characteristic for the pharma and comparable industries. Especially for assembly industries there is still another, perhaps the last, optimization potential.

This is an effect of the different throughput times and innovation-cycles of the different components and of the different technologies employed in such a system product. Fig. 12 shows a typical situation like that. Thus we will use this picture to outline the basic idea behind this optimization strategy, we would like to call **logical paralleling**. The basic idea of this system optimization scheme is to **profit** as much as you can **from the differences in the respective throughput times** of each component technology in order to maximize the overall system development efficiency. This is achieved by treating each technology different.[vii]

The overall **system** development **throughput time** is in general **determined by its slowest developing component.** In the example motor development project sketched in Fig. 12, this is propulsion technology, e.g. the new combustion technology "space charge ignition". Now, which steps do we have to take to optimally organize such an endeavor?

In order to build on that "most promising" basis a new, clean, fuel efficient and smooth running engine, we have to develop on top of this new base motor a new ignition and a suitable fuel injection system too. This in turn has as a consequence the development of an appropriate control-SW system for this new motor control HW (mechatronics) employing and implementing our new motor control methodology (space charge combustion).

---

[vii]    We will exploit this property of assembly industry innovation pipelines described in the next chapters to restrict our discussions and the "real life" examples to a somewhat virtual company, which we would like to call "XX-Corporation" for simplicity.

XX-Corporation serves as a placeholder for the most difficult and the most general class of inno-problems and/or inno-pipes, the assembly industry product innovation pipeline. Therefor XX-Corporation is assumed to be an international multi-brand, multi-technology and multi system-product company.

It thus shows all the basic and most of the non-basic properties and problems of good and not so good inno-management. Due to this, it will serve us as an example to discuss and demonstrate the very essence of what are the consequences of our novel inno-mamangement theory and approach in real life. This is even more so, because we carefully selected the input data for each of our XX-Corporation examples out of the rich data pool we did have access to.

This data pool has been fed by our own quite substantial professional experience as well as by quite a few international benchmarks we performed or we did have direct access too. These benchmarks covered pharma, chemistry, electronics, IT, communication, aerospace and car industries over a time period from 1995 to 2003.

Thus XX-Corporation never stands for a single company only. It rather should and must be considered to be a typical member of an enterprise or an inno-pipe typical especially for the assembly industries as a whole!

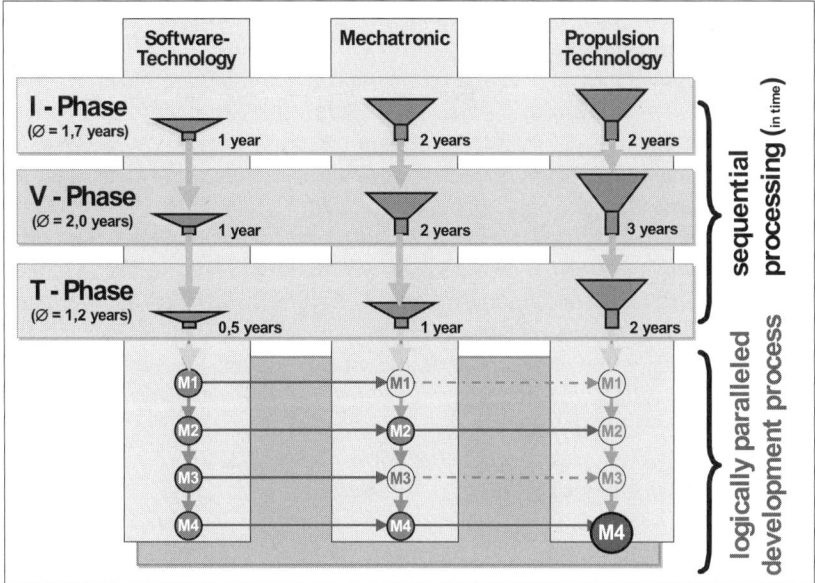

*Fig. 12: Logically paralleled system innovation/development projects and/or inno-pipes*

We now do ask you to remember the most widely known fact, that the average **throughput times** of these different technologies **are most different** too. **SW** usually has technology cycles of **some 2-3 years, mechatronics of some 3-5 years** and **combustion technologies** do develop much slower, let's assume a technology cycle time **TtM$_{comb}$ of about 5-7 years.** Whether or not these values are correct or realistic, is not really important for this example. It's their differences which do have an enormous impact on the respective system development efficiency, organization and control too.

From the **inno-gem** we could conclude, that **maintaining a constant speed** (of complexity reduction) in an inno-pipe is a decisive condition for its respective **optimal efficiency and efficacy.** This is only true relative to a assumed technology cycle time TtM$_{techj}$, cause "mathematically speaking" our model and its respective equations do **scale with their λ- or α-parameters.** As we know from their definitions (e.g. (D 12)) the "Time to Market" TtM$_{IP}$ of the respective inno-pipe IP is a decisive scaling parameter. This is nothing but the mathematical consequence of the fact, that **each technology** and its corresponding inno-pipe has **its own characteristic cycle time TtM$_{techj}$ with which** all the respective **processes and characteristics do scale!**

We can profit from this property of any (system) innovation-process, by exploiting the faster component cycle times to the benefit of a much higher system maturity. This is achieved by **doing more than one development cycle for a component technology for a certain maturity step of the overall system** (see (D 12)). Through this one achieves a **higher maturity with on average the same costs**. This is true, cause a component development team just sitting and waiting until a certain maturity level on the system side is achieved, does cost about the same amount of money than while working on its own targets.

Naturally there is a **price to pay** for these (potential) benefits. The price is a **substantial increase in the necessary persistency and in the maturity of the technical management** involved. It now has to be able to most clearly describe, determine and monitor the respective development and maturity targets of each component and of the system project at the same time. This is really **a challenge even for a** most well educated and **experienced "chief engineer"**. This is definitely **not an option for an "administrator"**, cause the quality of the technical evaluations and/or judgments of the different steps to take is the success determining part of this strategy.

## 2.8 Intermediate summary 2 – the inno-gem and innovation-pipeline design

**S2a)** There is **a fundamental inequation** (L 9) describing the **essential economic optimization properties of any** investment-, development and/or **innovation-process:**

$$(L\ 9) \qquad \overline{Ir}(IP,\Delta t) > \frac{\overline{C_F}(IP,\Delta t)}{\overline{EVA_S}(IP,\Delta t)} = \frac{\overline{C_F}(IF(IP,\Delta t))}{\overline{EVA_S}(IS(IP,\Delta t))}$$

    with $\overline{Ir}$ ( $IP$ , $\Delta t$ )     =    average innovation-rate of inno-pipe IP

    and $\overline{C_F}(IF(IP,\Delta t))$     =    average cost of a failing project from set IF

    and $\overline{EVA_S}(IS(IP,\Delta t))$ =    average Economic Value Added of an inno-success from set IS

    with respect to any arbitrary time-period $\Delta t$.

**S2b)** There are **just 3 basic strategies for the economic optimization** of any innovation-pipeline:

**1.)** **Maximize the innovation-rate Ir (IP, $\Delta t_k$)**

**2.)** **Minimize the failure costs** (costs of information Ci) **Ci = $C_F$(IP, $\Delta t_k$)** of an Inno-pipe IP

**3.)** **Maximize the** Economic Value Added **EVA$_S$(IP, $\Delta t_k$)** of the respective inno-successes IS(IP, $\Delta t_k$)

**S2c)** There are just **3 basic design options** how **to organize any innovation-pipeline** (see Fig. 8):

**1.) Cascading**     -    this is the basic design option to optimize and control any inno-pipe IP

**2.) Buy-in**     -    the predominant basic inno-pipe design option for assembly industries

**3.) Paralleling** -    a typical inno-pipe design option for the consumer goods manufacturers

Once the design rules (S2d and S2e) are respected, each of these 3 basic designs can be freely combined with any other option to form a suitable real inno-pipe design for the technology and/or market problem at hand.

**S2d)** To be at least capable to obtain an optimum inno-pipe efficiency $E_{IP}$ (see Fig. 9), each basic inno-pipe design (see S2c) must respect these 2 basic optimality conditions:

    **1.) Capacity-match (L 10):**

$$\sum_j C_{out}^{\,j}\,(IPstage\ (k)) \le \sum_i C_{in}^{\,i}\,(IPstage\ (k+1))$$

    **2.) Speed-match (L 11):**
    $$Ps\,(t_k)/Ps(t_k+\Delta t_k)\ =\ C_P\ \approx\ \ln\,(N_k/N_{k+\square})\,/\,\Delta t_k$$
    with               $N_{k+\Delta}$ = number_of_projects in the inno-pipe IP
                            at TtM = $t_k + \Delta t_k$

**S2e)** From the inno-gem we can derive 5 basic rules for the design of any inno-pipe IP:

    **1.) Never break the inno-chain nor allow to have weak links in between!** - this is the most basic and most cost saving, but, in general, the most violated design rule for any inno-pipe IP!

    **2.) Every stage of an inno-pipe can be optimized individually once the chaining conditions are respected** - this allows for quick, effective and easy optimizations!

    **3.) Always organize your inno-pipe according to its TtM- or maturity-levels** - this is most important for the development of optimized inno-pipe design and control strategies.

    **4.)** The **capacity strategy Ca(IP,TtM)** pursued **determines** the maximum **efficiency potential** of an inno-pipe IP and **Ca(IP,TtM) ≈ exp(-b\*TtM($t_i$)) is the optimal** one (Fig. 9)

    **5.) Logical paralleling** (see Fig. 12) is the **ultimate optimization strategy** once all the other options (S2a to S2e) are exhausted.

# 3.  THE MESOSCOPIC VIEW - CONTROL OF REAL WORLD INNOVATION PIPELINES

In chapter 2 we saw the enormous impact of the structure (see 2.3) and of the capacity design (see 2.4) on the respective maximum achievable inno-pipe efficiency. Naturally the values derived there are upper boundaries for real world inno-pipes which the pharma example (see 2.6) did demonstrate quite obviously.

Now we will demonstrate how the inno-gem  (see Fig. 6 on page 15) together with the application of classical control theory and the cost of information principle does allow for a much more efficient and stringent, thus more effective control of innovation-projects, -processes and -pipelines. The main task of any innovation management is to find a/the hopefully most optimal answer to question 6:

> *(Q 6)*  *When shall we deploy which kind and amount of resources for which topic in order to get an optimal short-, mid- and long term performance (output) of the part of the inno-pipe we are responsible for?*

This rather lengthy and complicated question summarizes two sets of underlying control problems:

> *(Q 6a)*  *How to execute an optimal temporal control of an inno-pipe*    or *how can you deal properly with the latency of an innovation pipe?*

> *(Q 6b)*  *How to execute an optimal topical control of an inno-pipe*    or *how can you optimize the success-function Ps(IP) of an inno-pipe?*

In the following chapters we will discuss these innovation control problems in detail and present a general purpose solution strategy, the "innovation phase control strategy" derived from the inno-gem.

## 3.1    Basic control properties of innovation pipelines

Fig. 13 and Fig. 14. illustrate the reasons, why it is so difficult to find appropriate answers to the questions (Q 6), (Q 6a) and (Q 6b) respectively. As can be seen in Fig. 13, the result of a/the basic management action item to control an innovation-pipe – the annual R&D-planning session - only shows up after some delay "Time to Market" (TtM). Thus you just have to decide on the respective R&D-projects in your portfolio on the basis of (R&D-success) expectations generated on experiences of past projects, technology- and market-developments. Due to this, any management action and/or R&D-planning executed on an inno-pipe is just doomed to result at least in some way in a closing of the feed back loop of the innovation pipe at hand. Cause of its respective delay-time characteristic (see Fig. 5 in chapter 1 or Fig. 10 in chapter 2) this produces a most pronounced tendency for (in general most harmful) oscillations of almost any innovation-pipe (see Fig. 15). Only some very few control strategies can handle this property, as we will see in the next paragraph 3.2.

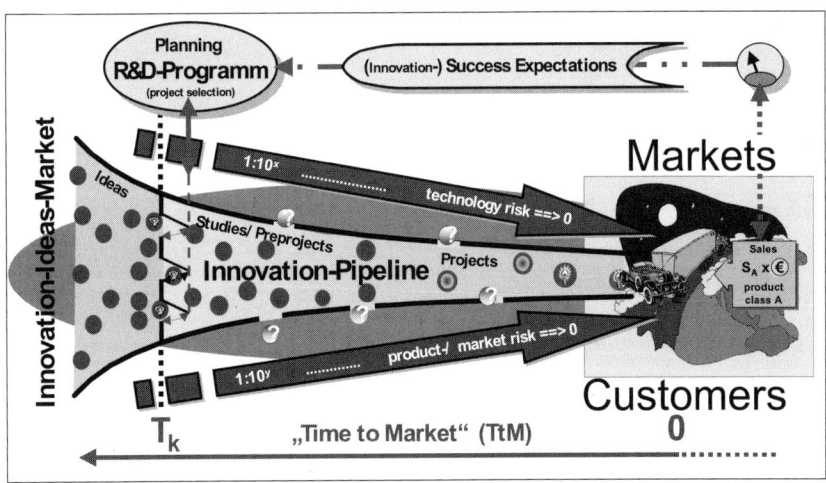

*Fig. 13: The „Nürnberg inno- funnel" - the management problem*

Let's suppose now we got that problem under control. Immediately we run into the next one. This is due to the same reason as before, our necessarily most limited knowledge on the future (outcome) of the projects in our inno-pipe (see Fig. 14). But there is some good news too. In spite of our limited knowledge of the future, we can use knowledge on the domain we are in, the technology we are using, the products/markets we are addressing and most of all we can use knowledge on previous similar projects to discern the "less good" from the "more good projects". Focusing resources on the latter ones helps to improve the success function Ps(IP) and thus the innovation pipe quite substantially. The basic principles to design a quality-gate and monitoring system will be outlined in paragraph 3.3.2.

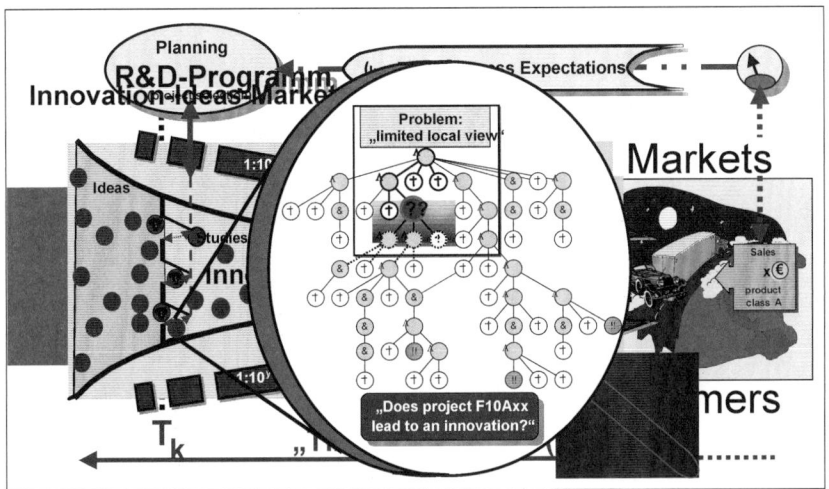

*Fig. 14:   The „Nürnberg inno-funnel" - the project-management problem*

## 3.2    Temporal control properties of innovation pipelines

As indicated before (see Fig. 13, Fig. 14) virtually any management action executed on an inno-pipe and any project-review, -selection and/or budgeting decision closes the feed back loop of a delay time system, which is the basic system control model for any inno-pipe (see chapter 1.3 and Fig. 5) according to the inno-gem (Fig. 6). Thus always remember the following lemma:

*Lemma 12 - the temporal instability of innovation-pipelines.*

*(L 12)   **Every management action executed on an inno-pipe influenced by information on its current status, performance, etc. has the tendency to produce harmful oscillations of and inside this inno-pipe.***

**Proof of Lemma 12- temporal instability of inno-pipes:**
    There are most elaborate descriptions on the oscillation tendencies and capabilities of fed back delay time systems in virtually any decent text book on control engineering. Thus we will refrain from reinventing the wheel and suppose Lemma 12 is accepted by any reader who accepts the inno-gem as a valid model for innovation pipelines and processes. Instead of that, Fig. 15 shows a plot of such an oscillation. There you can see the quite large real world R&D-system of our assumed XX-Corporation (see footnote VII, page 42) oscillating between classical extremities of any R&D-set up:
**Long term ivory tower research   versus   short term workshop research**

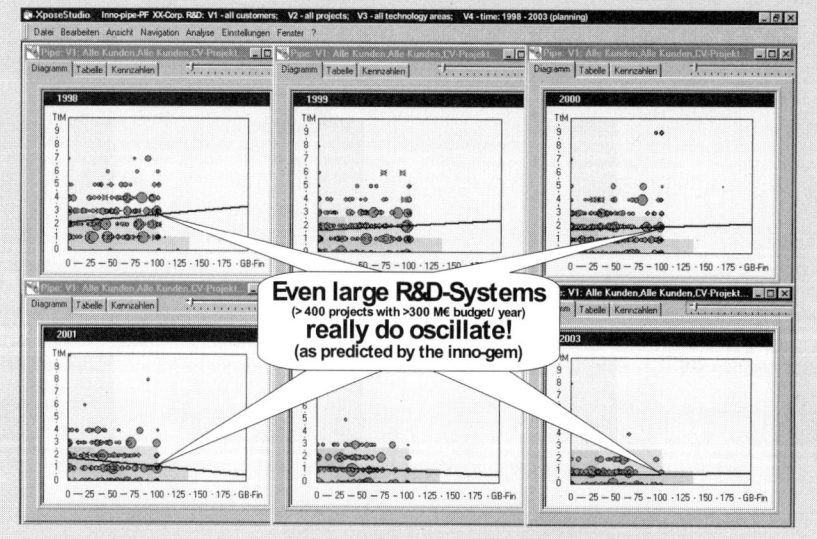

*Fig. 15: Ivory-tower versus a workshop-research oscillation of XX-Corp's R&D-department*

In Fig. 15 each quadrant represents the status of the inno-pipe of a year. The bubbles are the projects of the inno-pipe with their sizes proportional to the annual budget of the project. The y-axis is the estimated time to market (TtM) and the x-axis the funding part of the business units. The regression analysis lines in the annual project portfolios show the fluctuations of the funding behavior of XX-Corp. A positive slope corresponds to supremacy of business unit, more short term oriented funding. A negative slope vice versa corresponds to supremacy of corporate, more long term oriented project funding. Having several hundred projects annually in that R&D-pipe one can reasonably assume that the funding behavior does have impact on the kind and on the target of those projects and thus on the respective inno-pipe as a whole. **Q.E.D.**

Having accepted this tendency for instability inherent to any inno-pipe, we just have to ask ourselves, whether or not there are ways either to deal with or to circumvent that problem? Again control theory does render us appropriate solution strategies. To be more precise, there are essentially two approaches to deal with the inherent oscillation tendency of fed-back delay time systems:

**I)** **Predictor/estimator in the feed-back loop:**
Integrate an appropriate negative delay time element - a so called estimator/predictor - into the feed-back loop as sketched in Fig. 16 .

**II)** **Underlying control loops:**
Break up the feed-back loop into smaller subordinate control loops with acceptably small delay times and thus acceptably short oscillations and/or reaction time constants (see Fig. 17)

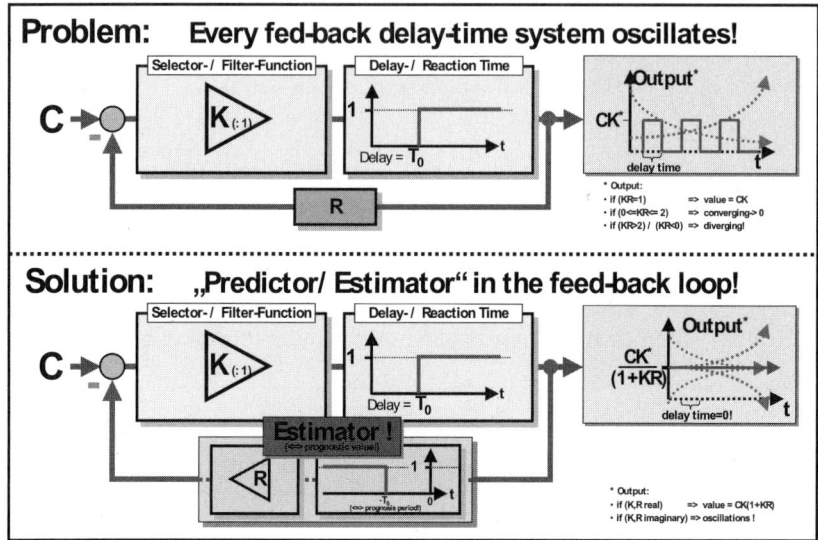

*Fig. 16: Solution strategy I for the oscillation problem of fed back delay time systems*

### To I) – Estimator/predictor in the feed back loop:

Mathematically speaking strategy I (predictor/estimator in the feedback loop) is the only appropriate solution to the oscillation problem at all. Any other approach (e.g. strategy II) is only able to ameliorate but not to solve the basic problem. The major disadvantage of this approach is to require exactly that which one in general does not have, the predictive knowledge.

The question to be answered is now, how we can construct or obtain the predictor/estimator required by the mathematically derived solution to the oscillation problem? We only do know two viable approaches to answer that question:

### Ia)  Inventor/entrepreneur control:

Integrate a most intelligent **"human estimator/predictor"** into your management feed-back loop and rely on his capability to do the right choices. This solution is nothing but the reestablishment of the classical "chief engineer function"[viii]. It works quite well, once you have got a good one, but it does not tell you where to get one nor how to recognize the one you need!

---

[viii] As we saw in the previous chapter 2.1 too, the inno-gem does formally derive quite a few of the most well known best practices in innovation management. This in turn fosters our trust in its validity.

**Ib) Innovation-path/-corridor control:**

- Statistically derive a **safe-operation area** for all the inno-projects in your pipe.
- Set-up a **monitoring system** which indicates when and in which direction an inno-project is leaving this safe operation area.
- Set-up a **project-portfolio management** system to keep the majority of your inno-projects inside this area
- Set-up a (very) **long term efficiency and efficacy controlling system** to ensure that the statistically derived safe-operation area is and remains properly defined

This is the technocratic approach to what an inventor/entrepreneur would do anyway. The major advantages of this approach are:

- It makes fluctuations/oscillations of the inno-pipe automatically visible (see e.g. Fig. 15) and thus controllable
- The monitoring and the portfolio management system can be highly automated (see below)
- It trains and helps R&D-Management to gain and keep insight in what is going on in the inno-pipe
- It gives management a good perception on how the inno-pipe will perform in the near future, at least within about 50% of the average product development cycle time (TtM).
- Due to that it should work decently even without well trained "chief engineers".

The price to pay for these advantages is a major effort in engineering and implementing as well as a minor effort in operating such a system and the according processes. How one can construct and operate such a system will be described in the chapters 3.4 and 3.5.

## To II) - Underlying control loops:

The basic idea here comes again from control theory. The idea is to make a system with a large delay time more responsive by subdividing it into a sequence of smaller and more responsive (less delayed) subsystems. As a side effect you get more information out of the system. This allows to apply even better control strategies and thus improves the situation further. This most well known approach is just transposed here from classical control applications to innovation management. A sketch of such an approach is given in Fig. 17.

A further nice feature of such an approach is the fact, that it can be integrated very easily into classical quality-gate control approaches. If one now adds some supporting innovation path control and the corresponding information and control systems, one easily can construct a most effective innovation phase control scheme and methodology. Details on how to do that will be discussed in chapter 3.4 and 3.5.

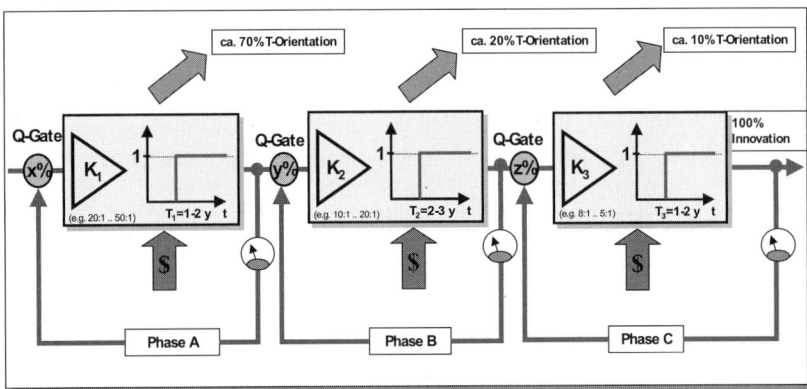

*Fig. 17: System-control loop for a large delay time inno-pipe using underlying control-loops*

## 3.3   Topical control properties of innovation pipelines – quality gate systems

The second part – the delay time element - of the basic control model of the inno-gem  (see Fig. 5 in chapter 1) was the reason for the temporal stability problems innovation pipelines do show. The first part of the basic control model - the selection or the success function (Ps(IP)) element - is the one that describes the filter characteristic of an inno-pipe. This is the term we want to optimize when answering question 6b:

> **(Q6b) How can we optimize the success-function Ps(IP) of an innovation-pipe IP?**

### 3.3.1   Results obtained from the inno-gem:

Before starting to set up topical control strategies, let us first remember which results and consequences of the inno-gem (see chapter 1 and 2) we could use so far:

a) **The fundamental inequation (L 9):**     $\overline{Ir}(IP, \Delta t) > \dfrac{\overline{C_F}(IP, \Delta t)}{EVA_S(IP, \Delta t)}$

b) **The two "golden rules" (see Rule 3) for inno-pipe optimization $S_l 1$ and $S_l 2$:**

   $S_l 1)$  Keep your inno-rate Ir(IP) as high as you can afford, but cut down on the necessary search costs CF(IP) as much as you can.

   $S_l 2)$  Keep the money saved by stopping "hopeless" projects in the inno-pipe and reinvest it in such a way that you maintain a stable flow of inno-projects in your filter.

c) **The inno-pipe success-function (D 4):**

   $$Ps(IP(t_i)) = \sum_k Ps(Ip_k(t_i)) * weight(Ip_k)$$

**Comment:**   The vector of the weights describes in principle the investment strategy for the inno-pipe. As an example, if the strategy is "each invested dollar has the same value", the weight $(Ip_k)$ is directly proportional to the invested budget in the project $Ip_k$.

d) **The quality gate definition (D 6):**

   $Qg_i(Ip_j, t_i) \Leftrightarrow Ps_i - \varepsilon < Ps(Ip_j, ti) < Ps_i + \varepsilon$

   where each $Ps_i = Ps\,(T_i\ \&\ M_i)$, with $0 \le i \le n$, is an element of an ordered set of values $\quad Ps_i = \in \{Ps_0, ..., Ps_n\}, \varepsilon > 0$
   and the $Ps_i$ have the properties $\quad Ps_i < Ps_{i+1}$ and $Ps_n = 1$.

**e)  The Bayesian calculus rules for the inno-success probabilities Ps (L 7):**

$$Ps(Ip_k) = Ps\,(Ts(Ip_k)) * Ps(Ms(Ip_k)\mid Ts(Ip_k)) \qquad \text{(T-push)}$$
$$= Ps(Ms(Ip_k)) * Ps(Ts(Ip_k)\mid Ms(Ip_k)) \qquad \text{(D-pull)}$$
$$= Ps(Ms(Ip_k)) * Ps(Ts(Ip_k)) \qquad \text{(incremental)}$$

**f)  The cost of information principle (L 6):**

$$Ci = \text{Costs (IF)} < \text{Profits (IS)} \quad \text{should be minimal!}$$

**g)  The capacity-match condition (L 10):**

$$\sum_j C_{out}^{\,j}\,(IPstage\ (k)) \le \sum_i C_{in}^{\,i}\,(IPstage\ (k+1))$$

**h)  The speed-match condition (L 11):**

$$Ps\,(t_k)/Ps(t_k+\Delta t_k) = C_P \approx \ln\,(N_k/N_{k+\Delta})\,/\,\Delta t_k$$

### 3.3.2  How to measure an inno-pipe's success – innovation quality gate systems

From the definition of innovation and of the inno-pipe success-function (see (D 4) and c above) we know that it is by definition an aggregation of a set of individual innovation project success-functions $Ps(Ip_k)$. Thus we can state

*Lemma 13 - the inno-quality gate system*

*(L 13) **Any linear metric based on an inno-quality gate system according to definition 4  is the best guess obtainable for the unknown success function Ps(IP) at the inno-pipe IP.***

> **Proof of Lemma 13:**
> As we know from definition 4 (see c) above) the success-function of an inno-pipe IP is the weighted sum of the respective values of the success-functions $Ps(Ip_k)$ of its projects $Ip_k \in IP$.
> At each gate $Qg_i$ the inno-quality gate system produces a set of $Ps_k$-values for all the projects in the pipe. Any linear combination of these $Ps_k$ –values thus is by definition a weighted sum of all the $Ps_k$-values at time point $t_i$ of all the projects $Ip_k$ of inno-pipe IP. This in turn is nothing but the value of the success function $Ps\,(IP(t_i))$ at time point $t_i$.  **Q.E.D**

We now can be sure, that a well defined inno-quality gate system renders us the best estimation of the success-function Ps of the inno-pipe IP, that we would like to optimize and control. Thus the next question raises immediately:

> *(Q 7)   **What are the necessary steps to build and operate a decent inno-quality gate system?***

Answering that question is not too difficult cause we can exploit the fact that a normal quality-gate system (QgS) for product development is just a subset of a corresponding inno-QgS. The methodology and the process are quite similar.

The major difference is that we expect an inno-QgS to deal properly with risk, better with the partially very low probabilities of success $Ps(Ip_k)$ of the individual inno-projects $Ip_k$ in the respective inno-pipe IP. Personally we are rather inclined to state that any inno-QgS has to be, better is risk management at a quite high level of sophistication.

*Rule 6 - the 6-step inno-quality gate system design rule*

**R6.1)** Formally **decompose a set of technology ($Ts_j$) and market ($Ms_j$) maturity criteria** by backtracking and relaxation of the respective requirements starting from customer- and/or market-readiness until an arbitrary start-risk or -chance $Ps_0 = (Ts_0 \& Ms_0)$.

The degree of fulfillment of both criteria ($Ts_j \& Ms_j$) define at any given time $t_j$ implicitly the values of the success-function $Ps(Ip_k(t_j)) = Ts_j(Ip_k) \& Ms_j (Ip_k)$ of any project $Ip_k$ of the inno-pipe IP at that time.

**R6.2)** Formally **define a set of - not necessarily equidistant - times $t_j$** starting at some arbitrary $t_0=TtM(IP)$ until a final time $t_N=0$ (introduction to the market, job1 or any equivalent point in time) with $t_j = t_{j-1}+\Delta t_j$ where $\Delta t_j$ is a parameter depending on j but not on the projects.

**R6.3)** Set up a **metric compliant with the Bayesian probability calculus rules for innovation-success** (see (L 7) or 3.3.1-e) above). This allows to calculate the individual project success-functions $Ps(Ip_k) = Ps(Ts(Ip_k)) * Ps(Ms(Ip_k) | Ts(Ip_k))$ etc. and the corresponding inno-pipe success-function $Ps(IP)$ as the weighted sum (see definition 4 or c above) of its individual project success-functions $Ps(Ip_k)$

**R6.4)** Integrate a **measurement metric** (weights, measurements and values) into the one of step 3 to account **for the compliance** of each inno-project $Ip_k$ **with the necessary time, cost and budget restrictions** of the inno-pipe at hand.

**R6.5)** Set up and run a **monitoring and control process** (e.g. a red-flag system) to ensure that the inno-quality gates are always been taken serious by each inno-project $Ip_k$ of the inno-pipe. Additionally this will keep management always informed about status and performance of the inno-pipe.

**R6.6)** Set-up and run a **long term monitoring and control process to verify the validity of the technology, market and planning maturity criteria** (steps 1 to 4) by backtracking their relative importance to successes/failures of at least some past and finished inno-projects $Ip_k$.

Especially for steps 1, 2, 4 and 5 we can learn a lot from existing development quality gate processes like e.g. the MDS [ix] or the CDS [x]. Thus we need

---

[ix]    MDS: Mercedes Development System (see also footnote III on page 3)

[x]    CDS: Chrysler Development System

not reinvent the wheel, but we should most carefully keep in our mind not to forget nor to neglect an appropriate appreciation of risk in general and of the market risk in particular.

The basic internal structure of an inno-pipe QgS is sketched in Fig. 18 in comparison to that of an existing development-QgS. Looking at Fig. 18 we immediately see that the market chance/risk criteria are not covered very intensively by this (traditional) development-QgS. Having in mind the most complicated and time consuming product decision processes, on can understand why some development-QgS pay not too much attention to risk in general and to market chance/risk in particular. An inno-QgS just cannot afford to do so. Both, technology and market risk management, are at the very kernel of what we do expect it to perform at least reasonably well.

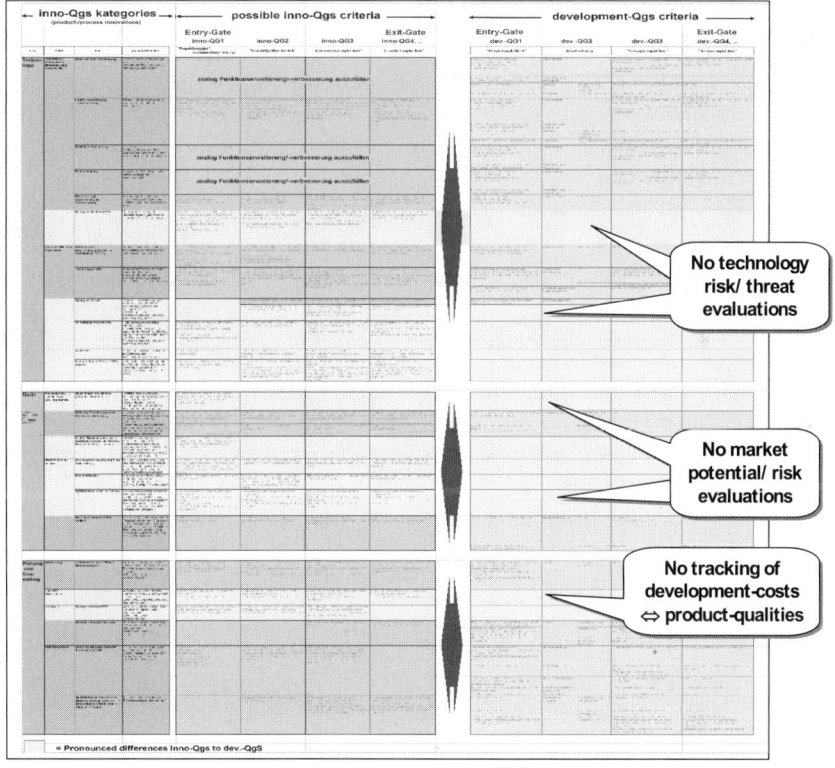

*Fig. 18: Scheme of an inno-QgS in comparison to existing development-QgS.*

There is a lot more to be said about how to construct and operate an inno-QgS in comparison with classical development-QgS. The major difference really is how it deals with risk/chance in general and with market risk/chance in particular. Remember the definition of inno-success and the inno-success function $Ps(Ip_k) = Ps(Ms(Ip_k)) * Ps(Ts(Ip_k))$ e.g. for incremental innovations (see (L 7) in chapter 1). Zero - e.g. for the market-chance $Ms_j$ of $Ip_k$ - multiplied with any value for its technological achievements $Ts_j$ definitely does not lead to any inno-success nor to decent Ps-values.

**We would like to finish that chapter with some concluding comments on the design of inno-QgS:**

a) The exact form of the individual quality-gate is by far less important than the logical and temporal consistency of the set of gates $Qg = \{Qg_0, ..., Qg_n\}$.

b) As sketched in Fig. 6 there is at least the possibility to enter a most productive learning cycle for the improvement of inno-QgS using the inno-gem (see chapter 1). The method is backtracking inno-successes (from set IS) and -failures (from set IF) to determine the decisive success-and/or failure-conditions and -criteria. Then use these to refocus the individual quality gates by applying an appropriate failure mode relaxation method.

   This learning from experience - a constant improvement policy - and the corresponding intensive analysis and discourse with the respective technical- and market-experts should be an integral part of any QgS!

c) The "time-step" $\Delta t_k$ between two gates $Qg_{k-1}$ and $Qg_k$ as well as the corresponding "maturity-step" together implicitly define the $\lambda$-value of the inno-pipe at hand (see Lemma 3 and its proof in chapter 1.1 and 1.2). The "maturity-step" is the required improvement in technological and market maturity proceeding from gate $Qg_{k-1}$ to gate $Qg_k$. It can be calculated by $\Delta Ps_k = Ps_k/Ps_{k-1}$.

d) The capacity- (L 10) and especially the speed-match criteria (L 11) require constant $\lambda$-values for any optimally designed and controlled inno-pipe. This holds from the start to the very end and for all sections of the inno-pipe (e.g. research-, predevelopment- and development-section) too.

e) The start-chance/risk $Ps_0 = e^{-\alpha}$, the technology-and/or product-cycle time $t_{pc}$, the "time-step" $\Delta t_k$ between two gates $Qg_{k-1}$ and $Qg_k$ and the corresponding "maturity-steps" $\Delta Ps_k = Ps_k/Ps_{k-1} = Cp$ do scale for optimally designed inno-pipes ($\lambda$=constant) according to following formula:

*Definition 13 - the scaling of the steps between two gates*

$$(D\ 13)\qquad \alpha * \Delta t_k = t_{pc} * \ln(\Delta Ps_k) = -\Delta t_k * \ln Ps_0$$

Thus we can derive the following scaling properties for time-step ($\Delta t_k$), maturity-steps ($\Delta Ps_k$), cycle-time ($t_{pc}$) and chance/risk ($Ps_0$) of any optimal inno-pipe the:

*Definition 14  -  the scaling properties of inno-QgS*

$$(D\ 14)\qquad \Delta t_k \quad \sim\ \textbf{\textit{ln}\ (}\textbf{\textit{$\Delta Ps_k$}}\textbf{\textit{)}} \quad \sim\ \textit{ln(maturity-step-size)}$$

$$\sim\ t_{pc} \qquad\qquad \sim\ \textit{technology-/product-cycle time}$$

$$\sim\ 1/\alpha \qquad\qquad \sim\ 1/ln(Ps_0)$$

$$\sim\ 1/ln(start\text{-}chance/risk)$$

Following these rules (from a to e) and steps 1 to 6 of Rule 6, it should not be too difficult to design a most effective inno-Qgs. Especially the sketched learning cycle (see b) above) is a most appealing feature of any inno-QgS designed according to this methodology. It allows for steady improvements due to a built-in learning from experience feature. On top of that the scaling properties (D 14) do render a solid basis to construct and run most efficiently (logically) paralleled system-innovation pipelines as described in chapter 2.5.2.

## 3.4   Dealing with latency – the inno-phase control approach

For an effective and efficient control of any inno-pipe IP we have to deal with its temporal stability problems (see 3.2) and the success-function optimization problem (see 3.3.2) at the same time. This is really most important cause the solution to the inno-pipe optimization problem is

- **not to do nor to find the right projects**, topics or technologies

but

- **to do the better bets on the more promising ones!**

Always remember that, as pointed out in detail in chapter 1 and especially in the chapters 2.2 and 2.3,

- **any innovation (project) necessarily is uncertain**

and thus

- **inno-pipe control really is sophisticated risk management at its very kernel.**

The inno-gem does render us a quite sophisticated inno-process model. Based on this model we can develop and evaluate appropriate betting strategies for innovations. The guiding principles for defining and evaluating these strategies are, just as derived for the inno-pipe design rules, the fundamental inequation (L 9), the cost of information principle (L 6), the capacity- (L 10) and the speed-match (L 11) condition in particular. Additional information on how inno-phase and inno-corridor control systems and the corresponding processes and metrics can be designed is described in [7].

**First we would like to sketch the basic idea behind the inno-phase control approach:**

a) An inno-pipe represents a dynamic equilibrium (see Fig. 13) between new, incoming project-ideas and inno-successes hitting their respective markets. For this (steady) flow of projects, it is a good idea to require that each part (phase or basic time-step) of the inno-pipe really does render the same relative contribution to the inno-pipes overall success and/or failure.

b) Assumed that, an optimal betting strategy is to invest into each phase proportional to its relative contribution to the overall success/failure.

c) Given a) and b) it is a wise policy to manage each phase such that its respective relative success-rate (the rate of projects reaching the next stage) remains at the same maximum level ($Ps_k$ = const. $0 \leq k \leq n$).

**Any inno-control scheme following a) to c)** automatically does fulfill the fundamental inequation, the speed- and the capacity-match conditions. It thus **leads to an optimal inno-pipe**, as pointed out in the chapters 2.1 and 2.2 and especially in chapter 2.3. In order to integrate the optimization rules for temporal (see 3.2) and for topical (see 3.3) control into an innovation-phase control methodology we have to perform the following 7 steps:

---

*Rule 7 – the 7 inno-phase control system design rules*

**R7.1)** Formally **define a set of innovation-phases Ph** $=\{Ph_1, \ldots, Ph_n\}$ where

    **a)** each phase $Ph_j$ is the time-period $\Delta t_j$ between two subsequent quality gates $Qg_{j-1}$ and $Qg_j$.

    **b)** within each phase each inno-project $Ip_k$ succeeds or fails to perform the corresponding maturity step $\Delta Ps_j$.

    **c)** every project $Ip_k$ that fails to perform the corresponding maturity-step $\Delta Ps_j$ is terminated.

    **d)** Every project $Ip_k$ within the phase $Ph_j$ has a well defined project-maturity and thus a corresponding value of its success-function $Ps_{j-1} < Ps(Ip_k) < Ps_j$.

**R7.2)** Formally **define a minimum/maximum speed of complexity-reduction** (e.g. a project survival rate) for all the inno-phases $Ph_j$ $\in\{Ph_1, \ldots, Ph_n\}$ in the pipe. Take special care to respect the scaling rules (D 14) as well as the speed- (L 11) and capacity-match (L 10) conditions while doing this step.

**Comment:** This is a most crucial step, cause here essential parts of your inno-pipe's efficiency and efficacy (the $\lambda$-values) are implicitly defined. It is most crucial to respect the scaling rules (e.g. $\Delta t_j \sim$ cycle-time $t_{pc}$) especially for multi-technology inno-pipes. These are rather common and virtually mandatory for system innovations.

**R7.3)** Formally **define a "safe operation area"** $A_{IS}=\{(\Delta t_1, \Delta Ps_1), \ldots (\Delta t_n, \Delta Ps_n)\}$ where each inno-project has to remain in during each phase (see Fig. 19). This safe operation area consists of a set of pairs of (preferably equidistant) time-steps $\Delta t_j$ and their corresponding maturity steps $\Delta Ps_j$. Carefully respect the statistics (experience gained through prior projects) and all the rules for optimal topical control (see chapter 3.3.2) derived, while doing this step.

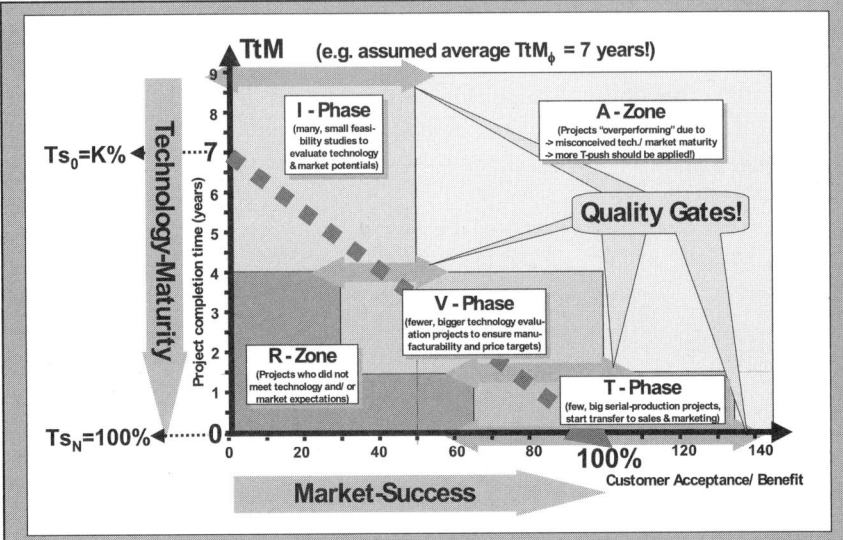

*Fig. 19: The innovation success-plane with inno-phases and quality gates sketched.*

**R7.4) Define an appropriate/desired general resource-deployment rule** for your set of inno-phases Ph. For each phase $Ph_j$ the sum of the resource-requirements of all projects $Ip_k$ in that phase may not exceed these limits (see e.g. chapter 5.7 and Rule 17). Again respect capacity-match (L 10) and capacity deployment strategies derived (see especially chapter 2.3) for optimality.

**Comment:** This step is exactly the point where business- and/or product-strategy must be integrated into inno-pipe design, its management and the innovation-strategies pursued. Cause an inno-pipe is a quasi-static entity with a life-time way beyond the ones of its inno-projects, the resource-deployment strategy determines the basic "betting strategy" (e.g. "technology-leader") of your company. Remember, not the projects, the betting strategy is the decisive item in the long run.

**R7.5) Design and run a** complete, persistent (to enable you to learn from experience) and as much as possible automated **capacity-use and inno-maturity monitoring system.** The set-up and the design of such an information system will be sketched in the next chapter 3.5.

**R7.6)** **Design and operate a sophisticated capacity-deployment and inno-maturity planning and monitoring process.** This process must at least cover the following topics:

    **a)** Planning and decision on project- and inno-phase capacity allowances (see chapter 5.7 and Rule 17).

    **b)** Monitoring, reviewing and termination of projects not meeting the standards (time, costs, inno-maturity/-quality).

    **c)** Reassignment of the resources unleashed in step b) within a phase or across phases.

    **d)** Monitoring/optimization of the flow of inno-projects through the pipe (fulfillment of speed- and capacity-match criteria).

**R7.7)** **Design and operate a (very) long term monitoring system to steadily improve the inno-phase control system** (learning cycle of the inno-gem) and processes in order to maximize the inno-success chances in the long run. The back-tracking and relaxation methods for inno-successes/-failures described in chapter 3.3.2 are the preferred methods to perform this task. A well designed and operated (automated) inno-maturity monitoring system (see step 5) is an additional great help to do this job.

The phase-control approach (doing steps 1 to 7) does intimately integrate temporal and topical control for any inno-pipe. The definition of a "safe operation are" in the innovation-success plane (see Fig. 19) is the cornerstone of this approach:

- It implicitly **defines a most general inno-success plane** on which the overall performance of any inno-pipe can be visualized and monitored.

- It is additionally an implicit **multi-year status monitoring approach** too, cause it always shows the complete status of the inno-pipe at any time. (see also Fig. 15 and Fig. 19).

- It thus intimately **marries topical and temporal control** of an inno-pipe cause one

  - immediately **visualizes the topical performance/ success** of any project in the pipe.

  - can most comfortably **track**, especially by comparing multi-period snap-shots, **the temporal control properties** of the inno-pipe at hand (see Fig. 15 as an example).

  - can **easily decompose both analysis modes** - temporal and topical - according to any predefined structure of a technology-, project- or market-clustering of the inno-pipe and its components.

## 3.5   Basic design of an information system for innovation-phase control

Due to the intimate integration of topical and temporal control of this new innovation-phase control approach, we need a new type of business information system too. It has to readily supply all the processes and all the parts of our innovation business system with appropriate information to perform their respective tasks properly.

Therefore let us first spend some thoughts on what are the basis tasks and elements of a business system which is focused on innovation. The scheme of such a business system shown in figure 19 demonstrates that we just do have

- only **one basic business process** to be controlled and this is the innovation- or **the product-creation process itself.**

- only **4 really important parts/players** in our business system. These 4 most important parts/players are

  1) the **customers/markets**, let us call them **vector 1**,
  2) the **products/processes**, let us call them **vector 2**,
  3) the **technology suppliers**, let us call them **vector 3** and finally
  4) the **time to market** needed, let us call it **vector 4** for simplicity.

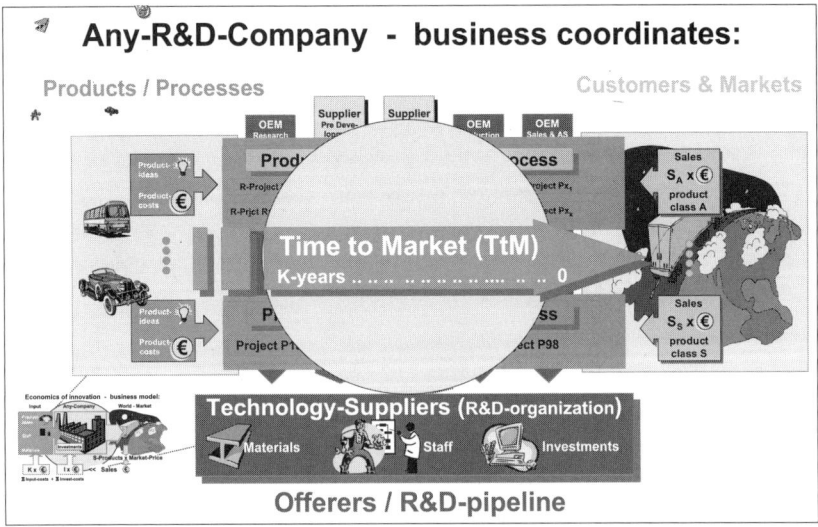

*Fig. 20: Sketch of a basic, most general (innovation) business system*

Let us now consider each of these 4 parts as the basic state space coordinates of our (innovation) business system. Having assumed that, we immediately realize, that using this assumption

- **any inno-control and -optimization problem translates into**
    - an optimal **path search problem** and into
    - a **state-space optimization problem.**

This corresponds to the problem of finding regions (4-dimensional cubes) of optimal performance within this very 4-dimensional state-space representation scheme of a business system.

Details of this new approach to set up and design business information systems and data warehouses are described in [8]. Thus and cause the focus of this paper is on innovation management and not on the design of (business) information systems, we will refrain from further discussing and outlining of this novel and most efficient approach to the design of business data warehouse systems.

One of the most beneficial features of this IT-system design is the fact that it allows most easily to design an almost completely automated system which is able to render all the information necessary to answer the key question (Q9) for virtually any inno-pipe control and optimization scheme conceivable (see Fig. 21).

---

*(Q 8)*  ***Who*** *does* ***what*** *for* ***whom*** *until* ***when*** *achieving which* ***result?***

---

As we can see from Fig. 21, a complete (inno-) project management scheme is the basis which enables us to render all the information necessary for inno-pipe control and optimization. Thus it is just impossible to do an optimized innovation management without a highly professional project management scheme and all the respective IT-systems supporting it.

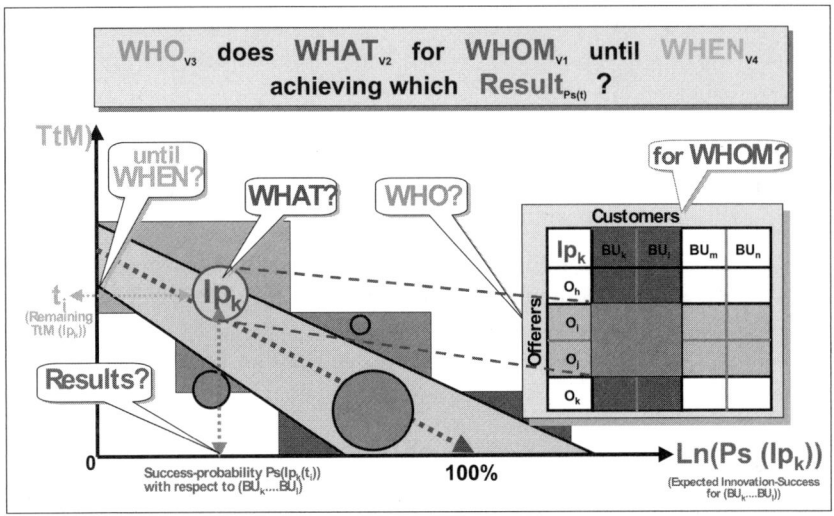

*Fig. 21: Scheme of an inno-pipe portfolio management tool*

In order to built an inno-pipe portfolio management tool we just have to map the stream of information coming from the normal, off-the-shelf project management IT-systems (SAP, MS-Project etc.) together with the information from the new inno-quality/-maturity monitoring system on to the inno-success plane described before (see Fig. 19 and Fig. 21). If we now are able to organize our business system such that

a) we introduce an **atomic information layer** structured according to our 4 basic business system-vectors v1 to v4, e.g. as atomic quadruples (v1, v2, v3, v4; $info_1$, ... , $info_N$)

b) **every business system-vector is in itself tree-structured** again, from the top down to the bottom, the atomic layer,

c) **all result values ($info_1$, ... ,$info_N$)**, like e.g. failure-costs, $EVA_S$, etc. **can be aggregated** very easily (e.g. like $EVA_S(IP) = \Sigma EVA_S(Ip_k)$),

then we can enter the most thrilling, highly automated and hierarchically structured innovation management and control analysis loop sketched in Fig. 22. Repeatedly going through this loop allows (innovation) management to analyze and track at any time most exactly what is going on in an inno-pipe. Innovation management and controlling can do this analysis with respect to

• the **costs** of the inno-successes (set IS) and/or inno-failures (set IF)

• the **profits or EVAs** of the inno-successes (set IS)

- **"Time-to Market"** of an arbitrary set of inno-projects $\{Ip_1, \ldots, Ip_m\}$

- **success- and/or failure-rate** for the inno-pipe or an arbitrary subset of phases or projects

- the **contribution of arbitrary parts of the (inno-) business system** to arbitrary subsets of its respective costs or EVAs

- **inno-project planning** and/or their more or less probable future results with respect to arbitrary subsets of projects by using sophisticated statistical analysis and projection tools and methods.

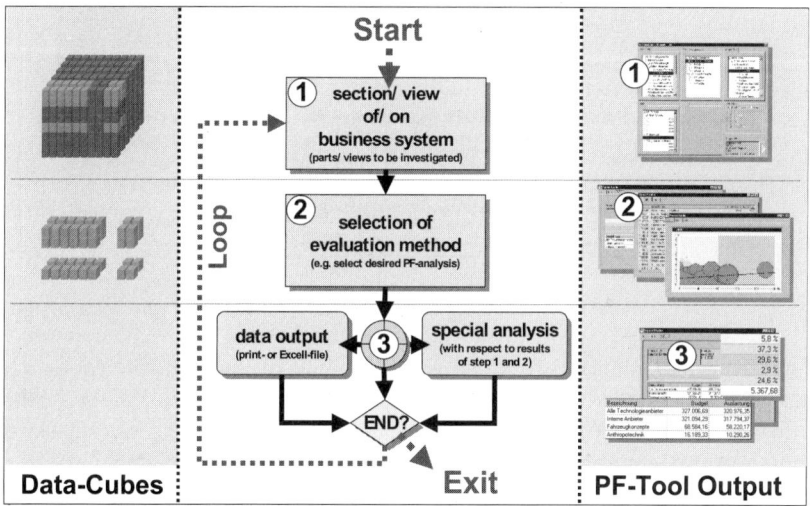

*Fig. 22: Scheme of a inno-business system analysis loop using the inno-pipe portfolio tool*

The possibilities for different hierarchically structured analysis modes of the (inno-) business system are virtually infinite using this most modular approach to the design of business data warehouses according to criteria a) to c). looking from the point of view of innovation management and control, the analysis of

- the **R&D-organization** corresponding to the (organizational) structure of the inno-pipe,

- the **innovation-pipeline** itself and of

- the **innovation-market**, where ideas for innovations are rated and traded, are of a special importance. Fig. 23 demonstrates how easily these features can be integrated into the proposed (inno-) business information system.

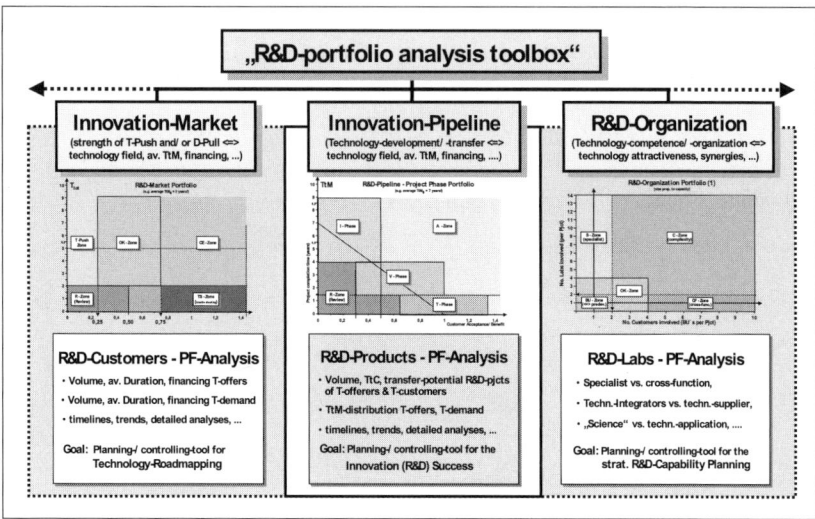

*Fig. 23: R&D-portfolio analysis toolbox for the proposed business data warehouse design*

Fig. 23 clearly indicates that this is not the end of the line. The (potential) analysis capabilities of the proposed IT-system design (see [8] for further information) are much bigger. Together with the proposed inno-phase or inno-corridor control methodology (see [7] for further information) from chapter 3.4, we are now in the most comfortable position to have a superior innovation management methodology - the inno-gem - and the corresponding tool-set to execute it appropriately available at the same time.

Thus everything is ready to enter a much higher level of sophistication in innovation management. The possibilities to excuse bad decisions with not having been able to know things better thus will diminish in the future.

## 3.6 The inno-phase control approach – application examples

In order to keep the more theoretical chapters 3.4 and 3.5 crisp and clear, we deliberately made them quite brief. This is especially true for chapter 3.5, where the design of the information system supporting inno-phase control has been described. To compensate for that, we now will give some selected examples on how an inno-phase control scheme could look like and how it might be executed. To make it more transparent to the reader we will pass through the examples following the general 3 step inno-business system analysis loop show in Fig. 22 on page 69 before. The steps demonstrated in this chapter are:

1.) **Selection of the view on the business system** – to demonstrate the possibilities of the business system navigator of the proposed information system design (see chapter 3.5).

2.) **Selection of the evaluation method** - here we will restrict the demo to an inno-pipe control scheme.

3.) **Special analysis** – here we will discuss especially a set of typical (individual) project histories.

**To 1.) - selection of the view on the business system:**

As described in chapter 3.5, the 4-vector design of the proposed business information system allows for an almost infinite number of different views on any business system. For maximum versatility, each business system should be tree-structured as shown in Fig. 24 below. This tree-structuring must be consistent until a lowest, the atomic level. This level determines the maximum granularity of any (possible) analysis to be carried out on the business system. It can be defined differently for each basic business system vector. In the example shown in Fig. 24 these granularity levels are reaching from

- all sales/markets down to the individual BU-customer of R&D for **vector V1 (customers)**,

- all R&D-work down to the individual R&D-project for **vector V2 (products/processes)**,

- the head of R&D down to individual groups/departments for **vector V3 (technology suppliers)**,

- all arbitrarily definable R&D-periods down to the individual year for **vector V4 (time)**.

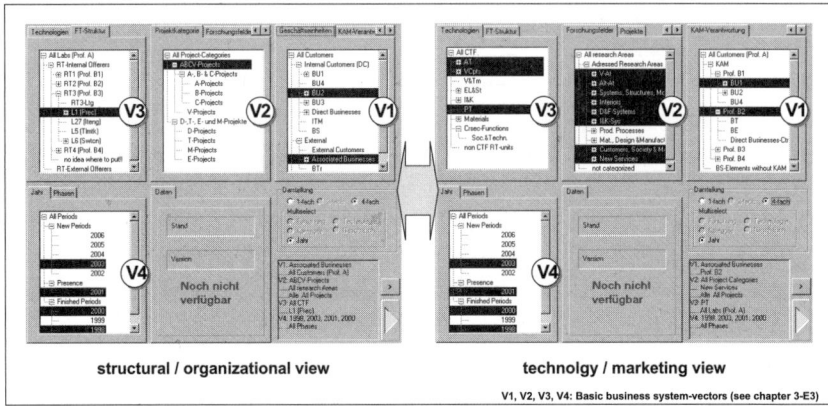

*Fig. 24: The 4-vector business system navigator with organizational and technical view*

Due to this most stringent design of the business system and of the corresponding information system, it is really easy to have multiple views on any business. This is shown in Fig. 24 too. As an example an organizational view and a technology-/marketing-view are sketched respectively. The organizational view is mainly dealing with departments, project-classes, business-units and so on. In contrast to that the technology-/marketing-view deals mainly with technology- and product-areas, key accounts and other comparable responsibilities.

Using this navigator it is most easy to select an appropriate view/part on/of any business system to be analyzed. This is only possible due to the atomic info-storage and structuring principle shown in Fig. 25. This principle allows the navigator to always select and hand over only the information relevant to the analysis at hand to the other parts of the analysis loop. These data-cubes (see Fig. 25) then form the "memory" on which any further analysis in the next steps described below will be based upon.

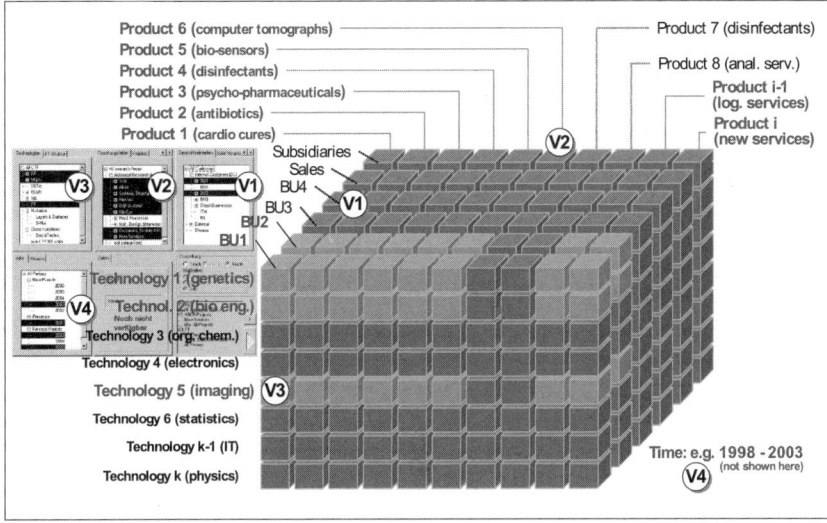

*Fig. 25: The selected navigator view and the corresponding 4-vector business system data-cubes*

In the description of the next steps of the business system analysis, we will see how these info-cubes allow to aggregate the relevant information according to the information needs of the analysis desired.

### To 2.) - selection of the evaluation method:

In chapter 3.5 we saw that there are quite a few different options to analyze any inno-business system using the proposed IT-Tool. Referring to Fig. 23 we will restrict the example evaluations presented here on the inno-pipe, better on the inno-phase control approach described in chapter 3.4. In this chapter we learned, that the introduction of

- **the inno-success plane** and of
- **the inno-phases** or the concept of a **"safe operating area"**.

as sketched in Fig. 19 is the cornerstone of this approach. We will see in the following figures 26 to 30 how, for the first time, this approach is able to integrate topical (see chapter 3.3) and temporal control (see chapter 3.2) possibilities for virtually any inno-pipe conceivable.

In these figures we see the green colored "safe operation" area –the inno-phases- and, colored red and yellow, the areas where your inno-projects better not be nor stay in. Thus this is the basic topical and, as we will see later, the basic temporal management control and action plane too. Just as with classical portfolios there are standard management strategies behind these areas. These strategies are:

$S_m1)$  All **projects within** each phase of the **green "safe operation area" do perform reasonably well**, as one can expect from past, similar project histories. In this case no special management action is needed (e.g. on the number of projects and their respective budgets).

$S_m2)$  The **capacity deployment** in **between the phases Ph$_j$** of the "safe operation area" should respect the "innovation strategy" of the firm set by management and the capacity- (L 10) and the speed-match (L 11) conditions for optimality. Thus
  - **always keep your inno-pipe in a steady state, never drain it!**
  - **once** too many **projects miss the next maturity step** (quality gate) **refill the pipe/phase** with new promising projects of corresponding maturity.

$S_m3)$  Once a **project gets and/or stays in the red "R-zone"** underneath the green "safe operation area" (see Fig.19) **it ought to be carefully reviewed with respect to** its
  - **overall desirability.** Is there really a reasonable expectation that there will be sufficient demand and buying power once the envisaged product hits the markets? If not do stop this "ivory tower R&D"!
  - **marketing policy.** If there is a market, what actions need to be taken to make it accessible for the envisaged product?
  - **necessary marketing investments.** Do the costs to get a reasonable market access and a sufficient market share endanger the overall profitability?

Always remember, an innovation is the coincidence of technological (Ts) and market success (Ms). In the "red-zone" there are mainly projects with severe market problems. Thus management is well advised to act on that and not to act on (perhaps) presumed technical problems instead!

$S_m4)$  If a **project gets and/or stays in the yellow "A-zone"** above the "safe operation area" (see Fig.19) one might be happy cause you have just got an over performing project doing (much) better than expected. Sadly this is not true in general. Instead of this positive conclusion, we have to **distinguish between the following 4 cases** to derive the appropriate actions to take once a project is in the "A-zone":
  1. **Technology- and market marketing/acceptance measurements are correct.** In this case we do have a clear demand-pull innova-

tion where technological maturity is behind schedule. Thus we should

- **invest** money and resources **to speed up technical development** until it is keeping up with the respective market maturity.
- **focus the management attention on technical items** cause they are the limiting factor in this case.

2. **Technology maturity is/has been over estimated.** Once you have a quality gate system installed this should not happen, thus

- do **review/revise your (technical) quality gate criteria** and measurements and check again.
- do **check whether this is a** (e.g. technologically difficult) **service task** and if this is the case, then try to delegate this task to others in order not to waste scarce and expensive R&D-resources.
- do **focus management attention on the quality gate process and on the R&D-project assignment process** to avoid this situation in the future.

3. **Market maturity/demand is/has been overestimated.** Again with a well run quality gate system, this should rather not happen. But, frankly speaking, especially for big market distances (long TtM), it is really difficult to estimate/measure the relative market potential of most future products. Thus,

- do **review/revise your (market) quality gate criteria** and measurements and check again.
- if you are still in the "A-zone" then proceed as described under the cases 1 or 2.
- do **focus management attention on** the respective evaluation processes for the (relative) **market potentials of your future products** (e.g. by establishing appropriate market-history and market-potential databases).

4. **Technology and market maturity are/have been over estimated,** then just redo your evaluation, check again and proceed according to the cases 1 to 3.

Having these 4 standard inno-phase control/management strategies (Sm1 to Sm4) in mind, let us now have a closer look on some real world inno-pipe examples to test the applicability of the approach presented. We would like to start with the inno-pipe for the process technology[xi] area T5 shown in

---

[xi] For a detailed discussion of the differences between product and process innovations/technologies please read chapter 4.2.3 starting on page 112.

Fig.26 to Fig. 28. These and the following project portfolios (Fig.26 to Fig. 30) are snapshots of parts of the R&D-pipe of our XX-Corporation, a large industrial R&D-department (see footnote VII, page 42). Thus one can expect quite some insight into what is going on inside a real world inno-pipe from these examples. They are definitely not academic! To better be able to interpret these figures (Fig. 26 to Fig. 30) we would like to give the following additional explanations:

- A **project** is represented by a **blue bubble** with its **size proportional to the annual budget** (resources) used.

- The **y-axis** represents the **"time to market"** measured in years as the time to completion for each project shown.

- The **x-axis** represents the **strength of the market demand**, measured as the fraction of the annual project budget financed directly by the customer or the future user of the R&D-result envisaged by the project. The other fraction of the project budget is financed out of corporate funds. Thus the relative part of the budget financed by the future user (customer) is a quite good measurement for the strength of his demand for the project and its respective (R&D-) results.

- The **dark line** is the **regression line of the relative budget financing** versus "time to market". It indicates the "spending policy" in an inno-pipe measured on a TtM-scale quite well!

**To 2.) - selection of the evaluation method – the technology area 5 inno-pipe example[xii]:**

Just looking at Fig. 26 we immediately see that the bubbles (projects) are lined up along an almost horizontal regression line with an average TtM of about 2 years. Most obviously this inno-pipe does not do its job properly, which is to identify and develop promising process technologies in tech-area 5. There are almost no new projects addressing future topics (5,6% of the total budget) and there is obviously not too much successful transfer of results to the customers (only 4,2% of the budget for the transfer phase) taking place either. Thus there is no or too little evaluation of future profitable topics and much too little transfer to customers and economic applications. It is not too difficult to imagine that this situation will rather aggravate than improve without proactive and consequent management action. This inno-pipe really needed quite some revitalization.

---

[xii]    The following examples shown Fig. 26 to Fig. 30 are taken from presentations on the IPMA conference in Berlin, 05. - 06.06.2002 and from presentations on the PM-Tage Wien, 27. – 30.11.2002

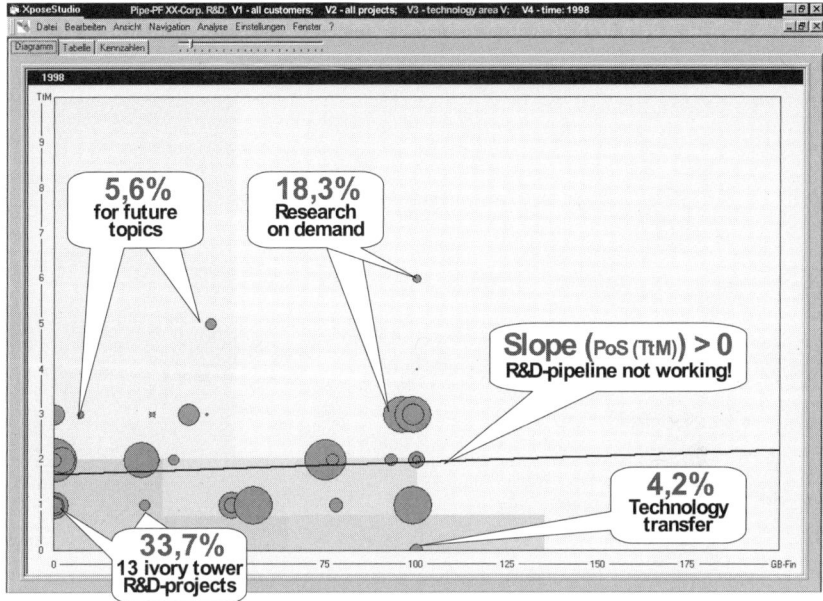

*Fig. 26: 1998 inno-phase portfolio for all R&D-projects of technology area 5*

How this could be done and how the corresponding result could look like is shown in the next picture (Fig. 27) underneath. This simulated improved inno-pipe example of tech-area 5 has been designed most easily using the proposed IT-tool (see chapter 3.5). We just had to do some reshuffling of the projects in the (real world) portfolio shown in Fig. 26 above. The actions simulated and their respective effects on the inno-pipe (Fig. 27) have been:

- **Scheduling** some **long term research** on demand **projects for transfer** to the business units.

- **Increasing investments and extending project duration** for some ivory tower (R-zone) and some research on demand (A-zone) projects.

- **Pushing the marketing activities** for some ivory tower (R-zone) projects.

These actions and their corresponding effects on the inno-pipe simulated show a most beneficial effect on the inno-pipe of tech-area 5. As the corresponding regression line indicates, the (simulated) inno-pipe now is working properly (see Fig. 27 below). It will stay that way, once management will continue to keep it in such a steady state. Consequent and stringent management action can achieve such a result in reality too. As an example we would like to show a 4 year snapshot of the very inno-pipe from tech-area 5 (see Fig. 28 below).

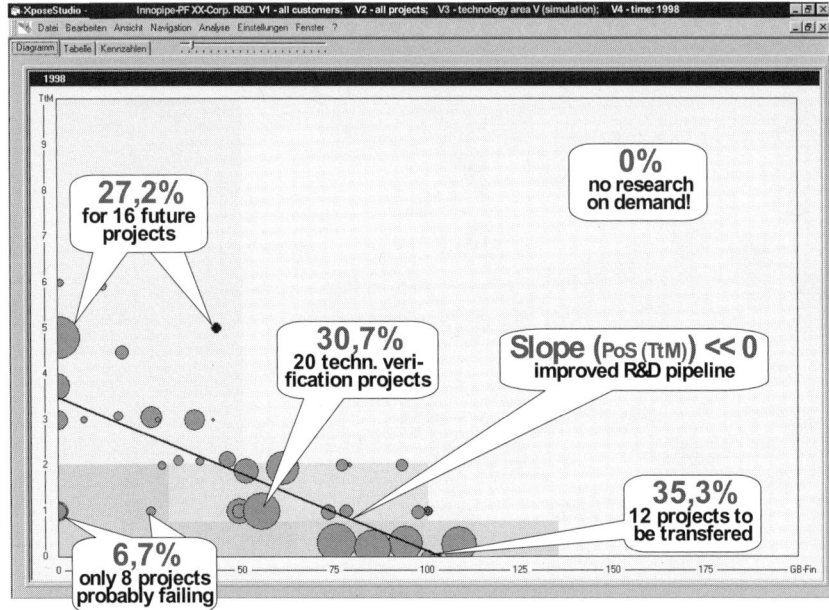

Fig. 27: Simulated improved inno-phase portfolio for technology area 5

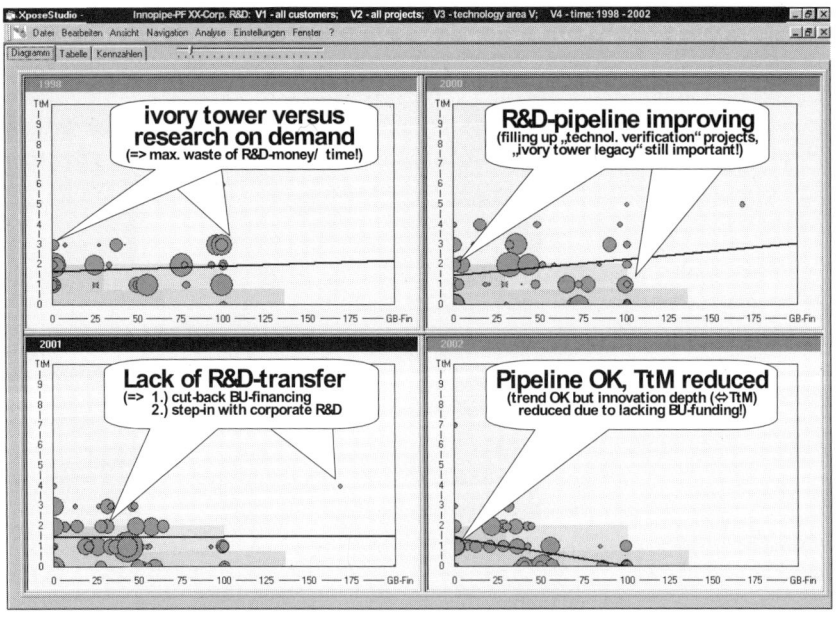

Fig. 28: 4-year snapshot of the real development of the inno-pipe of technology area 5

Fig. 28 demonstrates that one can obtain almost all the improvements achieved in the simulated example in reality too. Just look at the regression line of the year 2002 and how it improved since 1998. Naturally there are cut backs in real life all the times too. Most of the times they are due to disturbances coming from outside (of the pipe). Here a severe cut back in overall BU-financing lead to a shortening of the inno-pipe as a whole. One just could not afford that much long lasting R&D in that technology area anymore.

**To 2.) - selection of the evaluation method – temporal control of an inno-pipe:**

The green "safe operation area" - the inno-phases - does form the basis of the temporal control possibilities offered by the inno-phase control approach too. To demonstrate that, we would like to discuss the 6-year snapshot of the R&D-pipe of XX-Corp. (see Fig. 15 and footnote VII on page 42) again. As you can see in Fig. 29 below and in the previously shown Fig. 26 to Fig. 28 too, the inno-phase control approach most exactly shows you any tendency of your inno-pipe to oscillate. This is especially easy to detect, once a regression-line analysis is used. Temporal control now is really easy, even without having an experienced chief engineer available:

- Always **assign the projects** in your pipe **such that the regression line stays within the** boundaries of the respective **inno-phases** ("safe operation area").

Thus, by the very nature of its definition, the **inno-phases** are, better **do act as some sort of a predictor in the inno-phase control approach**. This is exactly what we need to safeguard temporal stability of the respective inno-pipe under control (see chapter 3.2 and Fig. 16 for further information)!

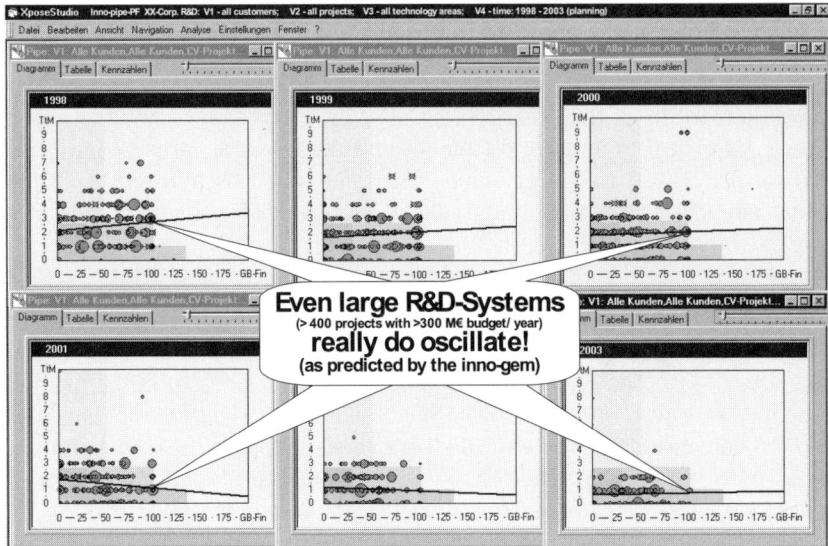

*Fig. 29: 6-year snapshot of a R&D-pipe oscillation of XX-Corp.*

## To 3.) – special analysis – some example analysis of real world project histories:

One of or perhaps the most appealing feature of the proposed business information system sketched in chapter 3.5 is the hierarchically structured zoom-in and zoom-out capability which it offers along any business system-vector. We will now apply this feature to individual R&D-projects and their corresponding project histories with respect to the inno-pipe they are or have been a part of. Before we will discuss the individual projects in more detail, we would like to explain first the grouping of the project history examples shown in Fig. 30 below:

- The **project histories A1 to C4 range from 1992 to 2005**. Thus some histories are in reality based on planning data. But planning is the best guess available for the future, isn't it so?

- The individual project **budgets** do range from **5 to 90 Mio. €** with an **average above 25 Mio. €.** Thus these are rather significant projects for our XX-Corporation and not academic examples.

- The projects A1 to A4 are examples of rather **typical ivory tower (A1, A2) or research on demand (A3, A4) R&D-project histories**. You better do not want to have too many projects of this kind in your inno-pipe.

- Each project history **B1 to B4** shows a rather **typical class of R&D problems**. It is most stunning, how long these problems can prevail without being solved or the projects being canceled in real world R&D pipelines.

- Projects **C1 to C4** are more or less the kind of **projects one would like to find in any reasonably well managed R&D pipeline**. Although only project history C4 is a really convincing example.

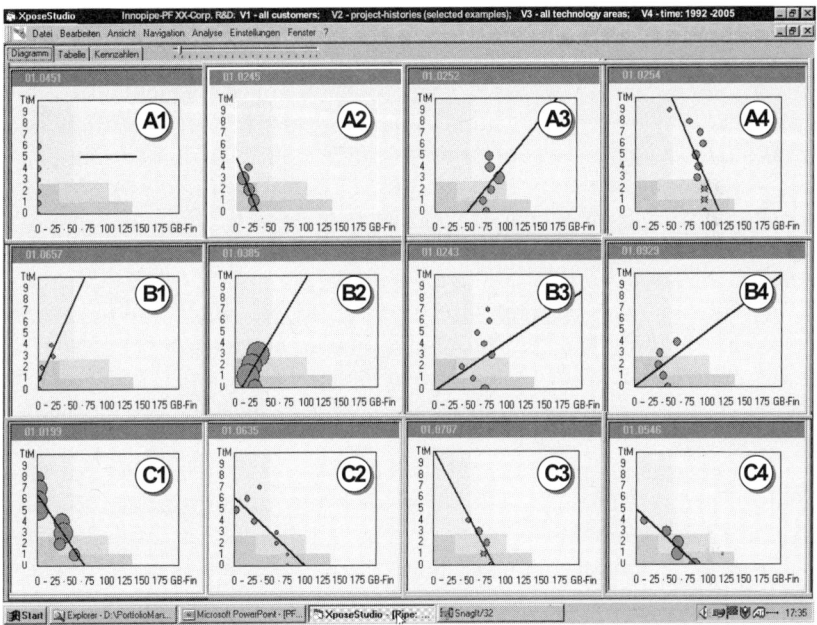

*Fig. 30: R&D-project histories as examples for typical R&D problem classes[xiii]*

Before discussing the projects presented a bit more in detail, we would very much like to remark that

- **none of the above projects really does comply with the capacity-match criteria (L 10).**

- **all the projects have been planned with a (quite) sub-optimal capacity deployment strategy.**

What we do see instead is a more or less constant budget scheme for all of the projects above. This makes life easy for clerks and the project planning

---

xiii See footnote VII on page 42; source presentations on IPMA conference, Berlin 06/2002 and on the PM-Tage, Wien 11/2002

staff but it is definitely sub-optimal for the respective efficiency and for the respective inno-success chance of any project. Here too the optimality conditions derived for inno-pipes (see chapter 2.3) do hold and thus should equally be respected.

To underline this we would very much like to remind the reader on the set-up rules for large military and aerospace projects. They too do start small (studies), grow once they pass their milestones and usually end big (deployment phase) or not at all. Perhaps there is something we could learn from?

**To project histories A1 to A4:**

**A1 and A2** are typical ivory tower research projects which did not find decent demand inside the organization for more than 6 years. **How long does it take** to convince an **organization to either stop or take a certain product/technology** is the only relevant question to be answered here? We personally do consider anything above 3 years to be at least questionable. If it takes longer, we think it is time to rethink your strategies and your decision making processes.

As far as projects **A3 and A4** are concerned, we would like to pose the question, **why did the organization not pour in more corporate resources** in the first place into such most wanted topics? One can reasonably assume that at least a little speeding up would have been most helpful for the projects and their customers too.

As a general comment to projects A1 to A4 we have to state, that these kinds of **projects histories with regression lines almost parallel to the y-axis are most common** for this very inno-pipe as well as for quite a few R&D-pipes in industry. A most thorough investigation done by [9] does support that too. Sadly **these project histories are all sub-optimal** from the point of view of a successful inno-management.

Obviously there is quite a strong tendency of R&D-project management once they found some customer acceptance to stick with it. The very principle of innovation calls for a steady improvement of the respective customer acceptance over time instead. This should show up in a regression line of the respective project history inside the inno-phases and not along the y-axis!

**To project histories B1 to B4:**

This is the **problem zone of projects not making progress** in general. The positive slope of the respective regression lines do show this most impressively. Thus one would expect to see intensive reviewing and redesigning activities in these projects, but we do not. Obviously there is some room for improvement inside the project monitoring processes, at least as far as these projects are concerned.

By strictly applying formal criteria (e.g. customer acceptance measurements) projects **B1 and B2 should have been canceled** immediately. They were not making progress as they are supposed to do. But again life is not that simple, as we can see in project history B2. This has been a most important and – most interestingly - a successful project too. But the **BU-customer** really needing and relying on the results of this project was just **not able (or willing) to finance it appropriately.** This has been the reason for this most problematic project history. We see in this example that in innovation management we always have to **put the organization** of the inno-pipe and/or the corporation **itself on the test too!**

As far as projects **B3 and B4** are concerned, one might be tempted to ask for the reasons of **the volatility of their respective customer acceptance.** Some more thorough reviewing, marketing and/or political engineering might have been helpful there.

**To project histories C1 to C4:**

Now we have reached the top performers. At least the general tendency to make (inno-) progress is Ok for all the examples in this group. But there is still quite some room for improvement:

- Project **C1**, a really heavy weight, did **start big and ended rather small.** This is definitely not what one would consider an optimal capacity design. Additionally its regression line is a little too much on the T-push side.

- Project **C2** is performing better with respect to that, but it shows some **difficulties** in homing into the inno-phase corridor **at the start.** On top of that it uses again an almost constant and thus a **sub-optimal capacity strategy.**

- Project **C3** is in contrast to C2 a little on the **demand-pull** side and it uses a **sub-optimal constant capacity strategy** too.

- Project **C4 is the best**. Its regression line is inside the corridor and the capacities tend to increase versus the end, the transfer phase where usually the project workload is increasing.

We really do hope that these typical but arbitrarily selected examples do help to explain the benefits of the proposed inno-phase control approach and of the corresponding IT-tool too. We are most certain that using this tool and this inno-control methodology could be a major step towards a better innovation management.

## 3.7    Intermediate summary 3 – the inno-gem and inno-pipe control

**S3a)**    **Innovation management and control** does have to **perform topical control** - the optimization of the inno-success function Ps(IP) - **and temporal control** (avoiding oscillations of/in the pipe) and the corresponding optimizations **at the same time.**

**S3b)**    **Every inno-pipe has an inherent tendency to oscillate (L 12)**, cause virtually every management action influenced by information on the current status of the inno-pipe closes a feed-back loop over its corresponding delay-time element TtM (see chapter 3.2 and Fig. 16).

**S3c)**    The inherent **temporal instability of any inno-pipe can be compensated by** either **introducing an "predictor/estimator"** into the feedback loop or by subdividing the inno-pipe into smaller, more responsive underlying control loops.

**S3d)**    The proposed **6-step inno-quality gate system (QgS)** is an optimal approximation for the unknown success-function Ps(IP) of any inno-pipe. It thus **is an optimal way to execute topical control** (L 13) on any inno-pipe (see 3.3).

**S3e)**    The proposed **7-step inno-phase control scheme integrates topical and temporal control** of an inno-pipe in an optimal way (see 3.4).

**S3f)**    The **universal inno-success plane allows to visualize and to execute topical and temporal control** on any inno-pipe in a most effective and efficient way (see Fig 19 and 21).

**S3g)**    The **scaling properties** (D 13) and (D 14) for the start-chance/risk $Ps_0 = e^{-\alpha}$, the technology- and/or product-cycle time $t_{pc}$, the "time-step" $\Delta t_k$ between two gates $Qg_{k-1}$ and $Qg_k$ and the corresponding "maturity-steps" $\Delta Ps_k = Ps_k / Ps_{k-1} = Cp$

(D 13)    $\alpha * \Delta t_k = t_{pc} * \ln(\Delta Ps_k) = -\Delta t_k * \ln Ps_0$

ought to be respected to guarantee for an optimized control and management of any inno-pipe.

**S3h)**    Applying the inno-gem does allow to install a most effective **learning cycle to steadily improve the inno-QgS as well as the inno-phase control scheme** by exploiting the experience (learning from prior project-successes and/or failures) generated inside the inno-pipe properly.

**S3i)** There is a **universal, 4-dimensional and most general (inno-) business system model together with a corresponding information architecture** to model and monitor inno business-transactions (see 3.5).

**S3j)** On the basis this 4-dimensional business system model one can design a most efficient and versatile **business data warehouse system to support** and highly automate all the tasks and processes of **any inno-phase control scheme** (see Fig. 22 and Fig. 23).

**S3k)** Keeping a (significant) **majority of the projects** and the respective regression **line inside the "safe operation area"** (the inno-phases) **integrates topical and temporal control** of the respective inno-pipe once the ino-phases are defined properly.

**S3l)** The capacity-match criteria (L 10) should be respected for individual R&D-projects too for optimal efficiency. Thus normally a **well planned inno-project starts small and ends big!**

**S3m)** The plot of the individual **project history in the inno-phase/-success plane** (see Fig. 30) **is a most useful indicator** of its respective inno-success chances and/or problems.

# 4.   THE MICROSCOPIC VIEW - CONTROL & MANAGEMENT OF INNO-PROJECTS

- "Each innovation and thus every inno-project just has to start from scratch again. The new, the unprecedented is the kernel of such an endeavor and thus, there is in general very little one can learn from others"

This or comparable opinions, e.g. like the ones focusing on the role of "creativity" which cannot be planned nor calculated in any innovation are most often used as a justification for taking unreasonable risks, for unforeseen failures and for reinventing the wheel over and over again. Although most widespread, this opinion is anything but true for innovation management. It is rather used by quite a few managers as a most comfortable excuse for not knowing and not wanting to learn things properly.

> **Remember**:   Generating **the "right idea" is definitively not the problem** in innovation management. **It is identifying** and selecting **the "more promising" ones** out of the many!

Thus just the opposite is at least closer to the truth. Experience, whether it is formally documented, e.g. in text books like this one, or it is in the brains of well trained "chief engineers" is to our opinion the most valuable resource at all in the innovation business. This experience, preferably assembled in one brain, comprises at best **equally profound knowledge** in the **technical,** the **economical and** in the **marketing domain.** Describing the linkage between these three fields of knowledge is a key focus of the inno-gem (see especially chapter 1) and the statistical filter approach described here. There is just a simple but most nasty problem with such a statistical approach when entering the single event, better the inno-project level:

- **You cannot measure, experience or evaluate chance nor risk on a single event/project basis, although chance/risk is there and it is definitively most relevant.**

How to deal with this problem for inno-projects and how to best exploit experience made in former, comparable endeavors will be the core of the following paragraphs describing the consequences of the macroscopic and the mesoscopic view of the inno-gem on individual innovation projects $Ip_k$.

## 4.1    The magic quadrangle of innovation - how to do the right things right

The **inno**vation **ma**gic **q**uadrangle, or the **inno-maq** for short, shown in Fig. 31 does intimately link

- the **product creation process** and

- the **demand/market creation process.**

This closely corresponds to the linkage of the technology success $Ts(Ip_k)$ and of the market success $Ms(Ip_k)$ for any inno-project $Ip_k$ at hand (see also Fig. 32) along its own TtM-scale. Additionally the inno-maq distinguishes between the

- planning-, control- or **invest-phase** (R&D-phase) and the

- cash-in, return- or **market-phase** (sales- and after sales phase) of an inno-project,

where the basic design decisions and success chances/risks (invest-phase) have been set and their respective positive or negative effects (market-phase) are being collected.

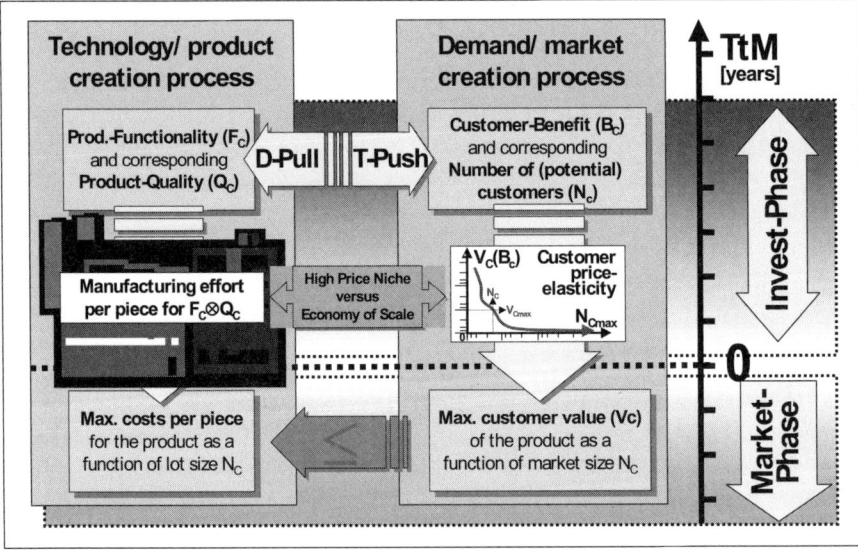

*Fig. 31: The innovation magic quadrangle (**inno-maq**)*

Once we rewrite the inno-maq in more mathematical terms, as shown in Fig. 32, we immediately see its intimate linkage to the inno-gem and its formulas, laws and strategies. To be more precise, we can describe the technology $Ts(Ip_k)$ and the market success $Ms(Ip_k)$ of any inno-project $Ip_k$ as described in the Definition 15 and the Definition 16:

*Definition 15 - the technology success $Ts(Ip_k)$*

*The technology success $Ts(Ip_k)$ of an inno-project $Ip_k$ is the probability $P_T$ to be able to deliver a customer functionality $F_c$ in a customer accepted quality $Q_C$ corresponding to a desired acceptable customer benefit $B_C$ for a suffi-cient large number of potential customers $N_C$ with the max-costs $C_{max}$.*

$$(D\ 15)\quad Ts(Ip_k) = P_T\ (F_C \otimes Q_C,\ B_C,\ N_C,\ C_{max})$$

*Definition 16 - the market success $Ms(Ip_k)$*

*The market success $Ms(Ip_k)$ is the probability $P_M$ to create for any given product functionality $F_c$ and a corresponding product quality $Q_C$ a suffi-ciently large customer benefit $B_C$ for a sufficiently large number of potential customers $N_{Cmax}$ to achieve a customer value $V_{Cmax}$ (corresponding to a maximum achievable market price), which is high enough to pay for all the product costs.*

$$(D\ 16)\quad Ms(Ip_k) = P_M\ (B_C \otimes N_C,\ F_C,\ Q_C,\ V_{Cmax})$$

Using these definitions, one immediately sees the dependencies (see also Lemma 7 in chapter 1) of the inno-success probability $Ps(Ip_k)$ from the tech-nology success probability $Ts(Ip_k) = P_T$ and the market success probability $Ms(Ip_k) = P_M$ of any inno-project $Ip_k$. These dependencies can even be de-rived formally, if one applies the rules of the Bayesian calculus to the three basic inno-strategies incremental, T-push or D-pull innovation.

*Lemma 7 – the 3 basic innovation strategies (reformulated):*

$Ps(Ip_k) = Ps(Ts)*Ps(Ms)$
$= P_T(F_C \otimes Q_C,\ B_C,\ N_C,\ C_{max}) * P_M(B_C \otimes N_C,\ F_C,\ Q_C,\ V_{Cmax})$     *(increm.)*

$= Ps(Ts)*Ps(Ms\ |\ Ts)$
$= P_T(F_C \otimes Q_C,\ B_C,\ N_C,\ C_{max}) * P_M(B_C \otimes N_C(F_C,\ Q_C),\ V_{Cmax})$     *(T-push)*

$= Ps(Ts\ |\ Ms)*Ps(Ms)$
$= P_T(F_C \otimes Q_C(B_C,\ N_C),\ C_{max}) * P_M(B_C \otimes N_C,\ F_C,\ Q_C,\ V_{Cmax})$     *(D-pull)*

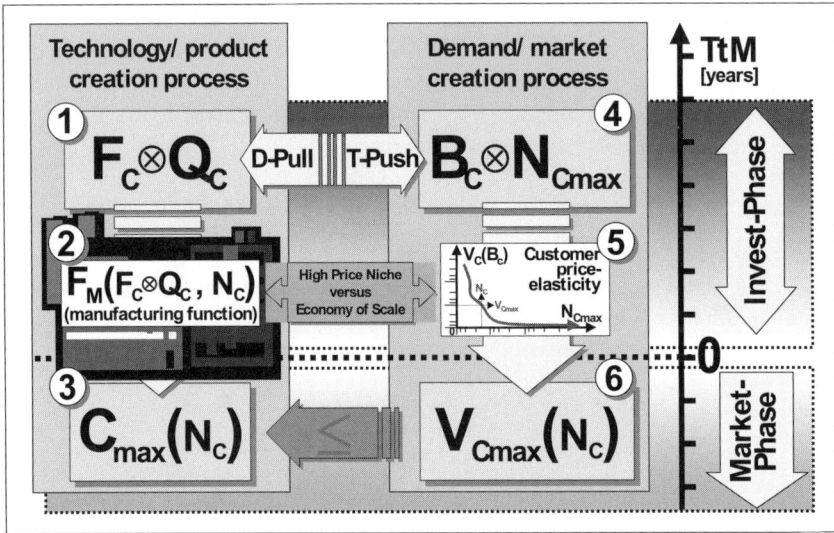

*Fig. 32: The inno-maq and the corresponding economic interdependencies[xiv]*

Thus the inno-maq does help us to better understand the close linkage between the product and/or technology development and the market and/or demand creation processes described by the conditional probabilities Ps(Ms | Ts) and Ps(Ts | Ms) in Lemma 7. In chapter 4.2 we will describe in more details on how this linkage and the knowledge about it greatly influences the relative success chances/risks of any inno-project $Ip_k$. Looking a little more into the inno-maq of Fig. 32 we see, that the link between the more conceptual/planning part of the invest-phase and the market-phase is done by

- the **manufacturing function** $F_M(F_C \otimes Q_C, N_C) = C_{max}$ for the product creation part and by

- the **price-elasticity function** $V_C(B_C \otimes N_{Cmax}, N_C) = V_{Cmax}$ for the demand/market creation part of any innovation process.

The manufacturing function $F_M$ does link the product features, the customer functionality $F_C$ and quality $Q_C$ via the market size $N_C$ (economy of scale) and the corresponding production technology to the maximal costs necessary and thus to the minimal price per piece $C_{max}$ to be economically feasible.

The price elasticity function $V_C$ does link the customer benefit $B_C$ - the solution independent, non-technical description of the customer relevance of

---

[xiv]  $F_C$= product functionality, $Q_C$ = product quality, $B_C$= customer benefit, $Nc_{max}$ = max achievable market size, $V_C$= achievable customer price (for $B_C$), $F_M$= manufacturing function, $C_{max}(N_C)$= maximum costs allowable for the market size $N_C$

the key product-features $F_C$ and $Q_C$ - together with a rough evaluation of the respective maximum market-size $N_{Cmax}$ to its respective maximum customer value $V_{Cmax}$ (maximum price obtainable) for any given market size $N_C$.

The determination of these two functions $F_M(F_C, Q_C, Nc)$ and $V_C(B_C, N_C)$ is at least in economical terms the decisive step in the innovation process. It is much too often not being performed with the appropriate care. This is true either in the product or in the market creation part and sometimes even in both parts of the innovation process. There you will find very often the key reason for quite some most expensively failing inno-projects $Ip_k$.

To summarize our inspection of the magic quadrangle, we have to state that any successful inno-project $Ip_k$ has to perform at least these 6 strategy- and interdependent steps to ensure an inno-success or a shut-down at minimal costs. Both is economically speaking a success in maximizing profits/minimizing losses:

---

*Rule 8 - the 6 most basic steps to take for an inno-project success*

**R8.1)**  **Define** preferably in non-technical, solution independent terms, corresponding to step 4, **the customer functionality $F_c$** and the **respective quality feature $Q_C$** of the envisaged product.

**R8.2)**  **Define** and evaluate an appropriate **manufacturing function $F_M(F_C, Q_C, N_C)$** for the envisaged market $N_C$ of the planned product.

**R8.3)**  **Derive** from step 2 a **maximum cost per piece $C_{max}$** you can afford to have. It **corresponds** to the minimal price or **customer evaluation $V_C$** you must achieve with the envisaged product.

**R8.4)**  Define and **evaluate**, preferably in solution independent terms, a detailed description of **a customer need/benefit $B_C$** and the **corresponding maximum market size $N_{Cmax}$** achievable according to the product definition of step 1.

**R8.5)**  Define and **evaluate** the **price-elasticity curve $V_C(B_C, N_C)$** for the envisaged customer need/benefit $B_C$ described in step 4.

**R8.6)**  **Derive** from step 5 a **suitable customer value $V_{Cmax}$** and the corresponding **market niche $N_C$.** $V_{Cmax}$ and $N_C$ do correspond to the maximum sales you can achieve with the product.

---

To ensure economic success $V_{Cmax}$ should be substantially higher than your maximum manufacturing costs $C_{max}$ and your market niche $N_C$ should be large enough to host you and your (potential) competitors in an economically reasonable manor.

The sequence, the relative importance and the difficulty to perform each of these 6 steps to innovation success is very much dependent on the innovation strategy pursued. This will be analyzed in detail in the following paragraphs.

## 4.2 The correlation of individual project success patterns within inno-pipes

From Rule 8 we know that there are 6 essential steps any inno-project has to take to be successful. These steps correspond to the cornerstones of the inno-maq (Fig. 31 and Fig. 32) introduced in the previous chapter. The main issue of this chapter is to describe in more detail following facts:

- These 6 steps to inno-success (Rule 8) are interdependent.

- These 6 steps have to be carried out in the right sequence, depending on the pursued inno-strategy T-push, D-pull or incremental.

The reason for this interdependency are the conditioned success probabilities $P(Ts \mid Ms)$ for D-pull and $P(Ms \mid Ts)$ for T-push inno-strategies in Lemma 7 describing the respective inno-success chances.

### 4.2.1 T-push inno-strategies and their peculiarities

We will now start to discuss these interdependencies with the supposedly most well known T-push innovations. By walking through this example sketched in Fig. 33, we will show the validity of our "complicated inno-mathematics" from chapter one, as well as we will demonstrate that these supposedly well known and described inno-strategies are not really understood completely up to now.

In the T-push inno-mode we start with a technology and/or a product idea at step 1 as shown in following figure. The normal and expected thing to do in a T-push mode is to finalize product development (step 1) until the ability to manufacture it reliably is established (step 2). After that the envisaged product does hit the markets at step 3 to look whether or not it can find a sufficient demand there (steps 5 and 6) to establish an inno-success. Just a short glance on Fig. 33 does show us that

- **any such supposedly "classical T-push innovation strategy" is not really a T-push one but it is in reality an irresponsible gambling with the economic odds against you!**

The reason for that is most simple and straight forward visible in Fig. 33. This strategy does perform almost all the investments necessary in steps 1 and 2 without establishing that there is a sufficient demand (step 4 to 6) $N_C$ with a sufficient high customer evaluation $V_C(N_C)$ for the product to pay for its costs. But the respective **price target** $F_M (F_C \otimes Q_C, N_C) = C_{max} < V_C$ **must be high enough to justify the investments made** in the first place. Thus we have to pose the decisive T-push question (Q 9):

> ### (Q 9)   What is a sound T-push strategy?

The answer to this question (Q 9) is already shown in Fig. 33, cause there you can see, that

- **the real hard and important part in any T-push strategy is the solution of the respective "marketing problem" (steps 4 to 6).**

It is not the design nor the production (steps 1 and 2) of a new innovative product. This we could have derived too from Lemma 7 itself. Cause there we do have this success probability of a T-push strategy:

$$\textbf{Ps(Ip) = Ps (Ts) * Ps(Ms | Ts)} \qquad \textbf{(L 7)}$$

The really nasty problem in Lemma 7 always is the computation of the respective conditional probabilities (here Ps(Ms | Ts)), thus we are well advised to always try to do this calculation first. This corresponds to doing step 4 to 6 first and much before finalizing product design (step 1) and the design of the respective manufacturing capability (step 2), which, by experience, does represent the bulk of the investment necessary $Tc(I_k)$ for the supposed innovation $I_k$.

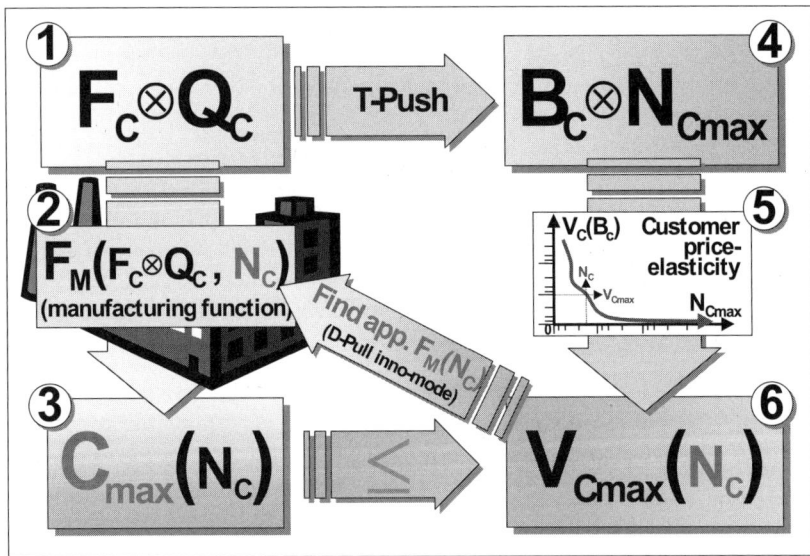

*Fig. 33: Walk-through a T-push inno-project loop[xv]*

---

[xv]   $F_C$= product functionality, $Q_C$ =product quality, $B_C$= customer benefit, $Nc_{max}$ = max achievable market size, $V_C$= achievable customer price (for $B_C$), $F_M$= manufacturing function, $C_{max}(N_C)$= maximum costs allowable for the market size $N_C$

This line of thinking directly leads us to the following rule to give an answer to the question (Q 9) posed above:

---

*Rule 9 - the 6 most basic steps to develop a sound T-push strategy*

**R9.1)** Start any T-push innovation strategy with **a minimal effort to evaluate and design** the product **functionality $F_C$ and** the corresponding product **quality features $Q_C$** of the envisaged product (step 1 in Fig. 33).

**R9.2)** **On the basis** of the results **of step 1 do**

   **a)** a **rough evaluation** of the **possibilities to manufacture** and the respective **expected costs per piece** of the envisaged product with the functionality $F_C$ and the quality $Q_C$.

   **b)** **describe** the envisaged **product-function $F_C$ and -quality $Q_C$ in** solution-independent **non technical terms** as a basis of the customer-benefit $B_C$ evaluation (step 4 in Fig 33).

**R9.3)** **On the basis of step 2 evaluate** an equivalent **customer benefit $B_C$** and the corresponding maximum **achievable market size $N_{Cmax}$** for the envisaged product with $F_C \otimes Q_C$

**R9.4)** **On the basis of** the customer-benefit description $B_C$ of step 3 do a price-sensitivity **analysis of the achievable customer price $V_C(B_C, N_C)$** as a function of the respective market size $N_C$

**R9.5)** **Select an appropriate (start-) market niche $N_C$ which offers a suitable customer evaluation $V_{Cmax}(N_C)$ to fit** with the achievable manufacturing costs $F_M(F_C \otimes Q_C, N_C)$

**R9.6)** **Finalize the product design** for $B_C(F_C \otimes Q_C)$ and **start the production** $F_M(F_C \otimes Q_C, N_C)$ for the selected market niche $N_C$ **with a cost $C_{max}(N_C) < V_{Cmax}(N_C)$** to achieve an innovation.

---

**Discussion of Rule 9 - Sound T-push strategies:**
   Rule 9 differs significantly from traditional T-push strategies, at least as we know them, with respect to the following items:

**a)** Rule 9 does demand for a **complete market creation/evaluation cycle** (steps 4 to 6 in Fig. 33) **before finalizing product design** and the start of the manufacturing.

**b)** Rule 9 does put the **bulk of the (verification) efforts on the market creation/evaluation process,** not on the necessary product creation efforts.

> **c)** Rule 9 does **assume the product creation effort to be much less complex** and uncertain than the corresponding market creation effort necessary (see b above).

To wrap-up the above said, we would very much like to state, that

- **any sound T-push strategy has to consider its respective market creation process to be the 80% problem to solve for success.**

- **technology is in general not the key question for such a strategy.**

Although we would rather consider the above said to be just good common thinking, it is in general not too well respected nor understood, at least to our personal experience. To illustrate these differences to traditional T-push management, we will now try to discuss and apply Rule 9 to a not quite hypothetical example, our example "fuel cell (FC) technology-push strategy".


### The example fuel cell T-push strategy[xvi]:

Let us assume our XX-corporation to have started a significant FC-R&D endeavor. XX-Corporation now is at a milestone, where different technical solution approaches have been evaluated. There are assumed 3 candidates with the following properties.

1) **PEM-FC with $H_2$-gas as fuel** – fairly mature (prototype), expensive (stack), bulky ($H_2$-storage)

2) **PEM-FC with $H_2$/methanol reforming** – moderately mature for prototypes (reforming technology), expensive (stack and reformer) but much less bulky (fuel storage)

3) **Ceramic-FC (SOFC or MCFC) with different liquid/gaseous fuels** – immature (unreliable lab-prototypes), less expensive and much less bulky (fuel storage, cooling)

Out of technical and time reasons, **XX-corporation decides to follow the $H_2$ PEM-FC route.** Now let us discuss, what the inno-gem and our "sound T-push strategy" would tell us starting with step 2:

---

[xvi]   Here we can narrow down the set of possible "XX-Corporations" to the major car-OEM´s GM, DaimlerChrysler, Ford, Honda and Toyota in particular, who, together with their respective suppliers (Aisin Seiki, BPS, …), pursue a fairly similar R&D-strategy betting on the PEM and on gaseous or liquid $H_2$ as the respective energy carrier.

**Step 2a:** - The price tag of some 1000 €/KW is extreme compared to the ICE[xvii] with some 20 €/KW.

**Step 2b:** - Quality standards (e.g. lifetime, operating temperatures, etc.) of the ICE are missed.
- Power, energy consumption and comfort requirement are met in general.
- Fuel ($H_2$-gas) is not available nor will it be available without legally required excessive efforts being made (see also [10] for additional information).

**Step 3:** - The maximum achievable customer benefit $B_C$ is about comparable to the one of the ICE.
- The maximum market size dropped by a factor of some 10.000 from some 50 Mio. vehicles per year to some 1000 vehicles/year due to the non-availability of $H_2$-gas as fuel. This does make only very limited fleet operations feasible (see e.g. [10] for additional information).

**Step 4:** - Extreme price-sensitivity due to the ICE competition with some 20 €/KW for mass- and special markets.
- Only an extremely small, politically motivated $H_2$-fleet-market is nevertheless possible (see e.g. [10] for additional information).

**Step 5:** - There is only an almost price-insensitive, perhaps a few 100 vehicle large market niche obtainable.

**Step 6:** - **There is no $H_2$ PEM-FC innovation due to an almost completely missing market.** This is due to the technologically based entrance decision to go for the $H_2$-PEM Option. The utmost which could be achieved is a "prototype customer research" endeavor, which does not at all justify any large investment!

At the end of the line our sound T-push strategy does render us quite a few items to most carefully think about:

- **The restriction on the single $H_2$- or PEM-route does endanger the mass-market capabilities for FC-vehicles significantly, at least for the next 10 to 20 years.**

- **The concentration on the single $H_2$- or PEM-route does increase the risk of a SOFC/MCFC vehicle technology invasion coming from power generation and home appliance applications.**

---

[xvii] ICE = Internal Combustion Engine

- **The linkage of the FC-vehicle technology to the availability of a sufficient $H_2$-infrastructure does intimately couple 2 most different markets and industries. This in turn increases market introduction and market penetration risks and problems dramatically.**

- **The USP of the FC-technology in the vehicle market is a very limited one and the price performance competition is one of the fiercest in industry there.**

Looking at this list of more or less unresolved questions, we would like to conclude, that XX-Corp.'s FC-endeavor has not yet tackled nor resolved its main T-push problem, the 80% marketing problem. There is still a very long way to go until this T-push innovation does achieve the necessary market acceptance. There should be made greater efforts to care about it too, cause investments already made and the ones being made increase steadily the pressure on the market specs to be fulfilled for an innovation success. To summarize the above and to finish this paragraph, we would very much like to state, that

- **good T-push and D-pull inno-management is a lot about posing the right questions in the right sequence!**

## 4.2.2   D-pull inno-strategies and their peculiarities

This class of innovation strategies is by inspection (from the inno-gem) and by experience the by far most successful one. There is an extremely long list of major technical and economical breakthroughs in this class of inno-strategies. The light bulb from T.A. Edison, some most important antibiotics like the sulfonamides, the $NH_3$-synthesis from Haber and Bosch and the light weight ICE[xviii] with G. Daimler as just one of its fathers are some very few selected examples of typical D-pull inno-successes. Having this quite impressive success-history in mind, we may ask ourselves, especially with respect to our results for the sound T-push strategies, the decisive question (Q 10) for this inno-strategy:

| *(Q 10)*   ***What is it, that makes D-pull innovations so successful?*** |
| --- |

To answer this question, it is a good idea to look at the major differences between these two strategies clearly visible once one compares Fig. 33 (T-push) with Fig. 34 (D-pull) below. The most obvious difference is, that in a D-pull inno-mode one does have at least have some appreciation of

- **the kind, the size and the price sensitivity of the customer problem to solve or the customer demand to satisfy respectively** (steps 4 and 5 in Fig. 34)

This is not common for T-push innovations. As a consequence, a D-pull innovation process is by nature focusing on the solution of the corresponding product and manufacturing technology development effort. To finalize this snap-shot comparison between D-pull and T-push innovation process we may say, that comparing Fig. 33 (T-push) with the situation sketched in Fig. 34 (D-pull)

- **a T-push innovation strategy has to focus on the marketing problem (step 4 to 6) while a D-pull innovation strategy has to concentrate on the technical problem (step 1 and 2).**[xix]

---

[xviii] **ICE** = Internal Combustion Engine
[xix] T.A. Edison made several thousand tests before having invented the D-Pull innovation electrical light bulb. In the pharma industry it is quite common to scan some 5000 to 10.000 substances before finding a cure to a targeted disease!

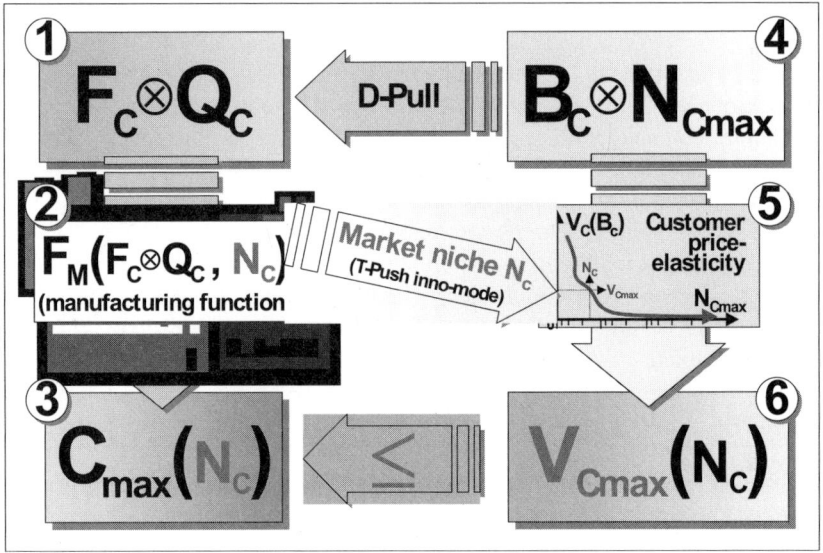

*Fig. 34: Walk-through a D-pull inno-project loop[xx]*

Keeping this in our mind, it is most bewildering to see that quite a few R&D-departments (inno-pipes) are designed and run like a D-pull inno-pipe focusing on the technical problem while they are operated under the regime of a T-push business strategy, not knowing neither the customer problem/demand nor the market size and elasticity properly. This mode of operation is just doomed to perform suboptimal, at least economically (see also Fig. 33, Fig. 34. To our personal experience this misalignment of inno- and business-strategy with the R&D-structure and their corresponding processes can be seen much too often in all kinds of industries. Keeping the above said in our mind, we do understand why D-pull innovations are in general the at least economically much more successful ones:

- **D-pull innovations in general need not to convince the customer nor the markets**

The market is already there. It is at least assumed to be there, right from the start. This is investors paradise, cause convincing customer/markets is in general a quite time consuming, a costly and a tedious will-forming process. Additionally the respective results are in general uncertain. This is the very reason why a T-push innovation just has to focus on this part (steps 4 to 6) in

---

[xx]   $F_C$= product functionality, $Q_C$ =product quality, $B_C$= customer benefit, $Nc_{max}$ = max achievable market size, $V_C$= achievable customer price (for $B_C$), $F_M$= manufacturing function, $C_{max}(N_C)$= maximum costs allowable for the market size $N_C$

particular. Having clearly understood the key difference between these two basic inno-strategies, we may now be able to give with the following rule 10 an answer to question (Q 10):

---

**Rule 10** - *the 6 most basic steps to develop a sound D-pull innovation strategy*

Starting from a decent knowledge of the customer demand $D_C$, the respective market size $N_{Cmax}$ and sensitivity $V_C(B_C)$ do:

**R10.1)** Design a **product functionality** $F_C$ and a corresponding **quality** $Q_C$ such that the envisaged product does **meet customer expectations $B_C$ for the planned market niche** $N_C$.

**R10.2)** Design and evaluate a **manufacturing capability** $F_M(F_C \otimes Q_C, N_C)$ for that product and for the addressed market size $N_C$.

**R10.3)** Check if the **achievable product costs per piece from step 2** $C_{max}(N_C)$ are sufficiently **smaller than the expected customer evaluation** $V_C(N_C)$.

**R10.4)** If this is the case, then **do a minimal market test** (steps 5 and 6 in Fig.34) to guarantee the results of step 3 and if this is true **finalize product design** (step 1), **manufacture and market** the product.

**R10.5)** If the **product costs per piece $C_{max}(N_C)$ are too high,** do a **serious price-elasticity analysis** (step 5 in Fig 34) to find an **appropriate market niche** $N_C$

**R10.6)** If **step 5 has been successful continue with step 3 else** either **stop** the innovation **endeavor or look for a production-process innovation** with $F_M(F_C \otimes Q_C, N_C) = C_{max} < V_C (N_C)$ to meet the marketing condition

---

Following these 6 most basic steps, it is ensured, that your D-pull innovation process does address the key-problems of this innovation strategy in the right sequence:

1) **Look for a decent product design that really meets customers (true) needs $B_C$.**

2) **Look for an appropriate manufacturing process to achieve an economically reasonable price target for the envisaged product and the selected market niche $N_C$.**

### 4.2.3   Incremental innovation strategies and their peculiarities

This is the most frequent and by the way the most easy to pursue innovation strategy. It is almost the only one possible to pursue in the assembly industries. E.g. a system-OEM almost by nature does have to incrementally develop his products and his corresponding USP's[xxi] via a sequence of component innovations and the subsequent system improvements. Thus we can easily answer this question:

> *(Q 11)   What are the key properties and success factors of incremental innovations?*

Just looking at Fig. 35 and recalling Lemma 7, stating that technology Ts and market success Ms are independent from one another in this case, we see that in the incremental innovation strategy

- **there is key knowledge on the product creation (step 1 to 3) as well as on the market creation process (step 4 to 6) available right from the start.**

So there is no big risk, neither technologically nor from the marketing side. The only crucial step is the

- **selection of a suitable market niche $N_C$ to ensure an economic success with $C_{max}(N_C) < V_C(N_C)$.**

In this case our so called "manufacturing function" $F_M(F_C \otimes Q_C, N_C)$ does play a decisive role. From an inspection of our inno-maq in Fig. 35, we should expect that process technologies become more and more important, once any product-differentiation strategy via a sufficiently large customer benefit $B_C = (F_C \otimes Q_C)$ approaches saturation. This is in correspondence to experience with mature markets and mature products as well. So there is not too much we can learn additionally by a deeper inspection of our inno-maq in this case, compared to the other two (T-push and D-pull) strategies discussed in detail in the former paragraphs.

Thus we would like to use the incremental innovation strategy to introduce and discern the two main classes of innovations:

1) The **product innovations** $B_C = (F_C \otimes Q_C)$ **creating a new** or redefining an improved product and/or **market of size $Nc_{max}$** (steps 1, 4; Fig 35).

2) The **process innovations** $C_{max} = F_M(F_C \otimes Q_C, N_C)$ **defining via** an economically feasible price $C_{max}$ **an accessible market share $N_C$** of an in essence already known (system) product/innovation (steps 2, 5; Fig.35).

---

[xxi]   **USP** = **U**nique **S**elling **P**roposition according to the economic theory of M. Porter

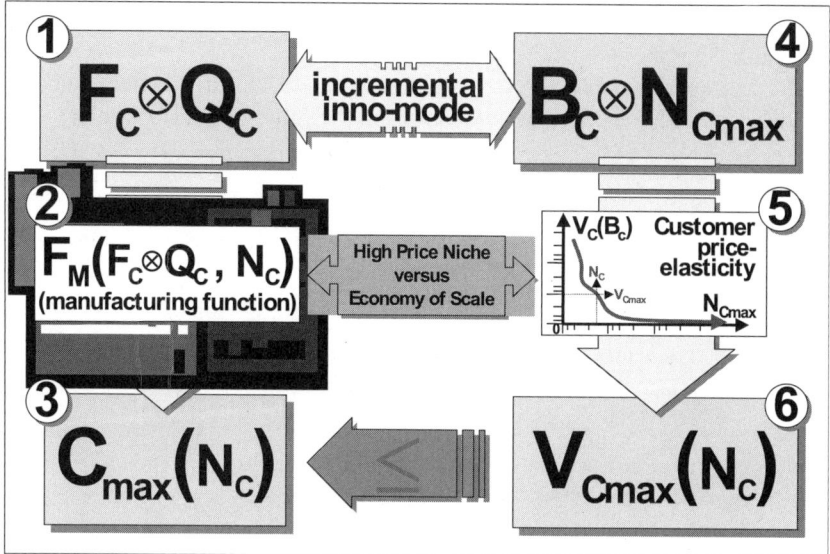

*Fig. 35: Walk through an incremental inno-project loop[xxii]*

It is most important to clearly distinguish between these two classes of innovations, cause

a)  for **product innovations** at least key parts of the **innovation chain** have to be in general an **integral  part of the respective firm** and/or inno-pipe, but

b)  for **process innovations** at least the **implementation and/or tool manufacturing** part just has to be **bought from a supplier.**

Therefore process innovations in general are more difficult to protect and to sustain than this is possible for a product feature innovation e.g. via a patent. If an enterprise is going for the process optimization option, it should be prepared to always keep on striving for process improvements. Competition very quickly will be able to challenge its achieved relative price/performance advantage.

There is a lot more which could be said about the respective properties of product- and process innovations, how they interact and how they should be planned and managed. Once one always respects those two key differences a) and b) mentioned above, then in general all the definitions, lemmas and

---

[xxii]  $F_C$= product functionality, $Q_C$ =product quality, $B_C$= customer benefit, $Nc_{max}$ = max achievable market size, $V_C$= achievable customer price (for $B_C$), $F_M$= manufacturing function, $C_{max}(N_C)$= maximum costs allowable for the market size $N_C$

rules derived so far, do apply equally with only minor modifications for both innovation classes.

To finalize this paragraph we would now like to summarize our results on how to deal with incremental innovations in the following 6 steps rule:

---

*Rule 11 - the 6 most basic steps to a sound incremental innovation strategy*

Starting with an equally quite good knowledge on the customer demand $B_C$, its respective maximum market size $Nc_{max}$ and on the realization chances of the corresponding technical features $F_C \otimes Q_C = B_C$ of the planned product do:

**R11.1)** Design **product functionality $F_C$ and quality $Q_C$** to meet **customer expectations $B_C$**

**R11.2)** In parallel with Step 1) **verify the corresponding market size $Nc_{max}$** and **evaluate** at least as a rough approximation its **respective price sensitivity $V_C(B_C)$**

**R11.3)** Based on the results of the product design in step 1 do as early as possible at least a rough design and calculation of the **corresponding manufacturing capability $F_M(F_C \otimes Q_C, N_C)$** to achieve a **rough estimate of the achievable costs per piece $C_{max}(N_C)$**

**R11.4)** Based on the evaluation of the costs per piece $C_{max}(N_C)$ from step 3 do a thorough analysis and comparison of the suitable **market niches $V_C(B_C, N_C) >> C_{max}(N_C)$** to identify and select an **appropriate and economically ($N_C$) feasible one.**

**R11.5)** If step 4 has been successful, finalize **product design $B_C = F_C \otimes Q_C$ and start production and marketing.**

**R11.6)** If step 4 has not been successful look for a possible, easy and quick to implement **process innovation for $F_M(F_C \otimes Q_C, N_C)$** to find a decent market niche $N_e$ and continue with step 4, **else stop the innovation** process immediately.

---

## 4.3   Basics of knowledge management – learning from failures/successes

Once we started our development of an integrated model of the innovation process (inno-gem and inno-maq ), we did not at all think of looking into knowledge management issues. It seemed to be far away, almost exotic judged from what we wanted and hoped to achieve with our endeavor. But the more we did try to understand the "logic of innovation" in order to find the most solid foundations obtainable for our theory and model, we learned that even the foundations of knowledge management are an integral part of it. To be more precise, they are a logical consequence of our statistical filter approach in general and of our proof tree logic paradigm in particular (see also Fig. 4, Fig. 6, the proofs of (L 2) and (L 3)). Numerous discussions and a long history of exchanging ideas between W.Klein (see also [5]) and us did create and foster our trust in this approach. We realized that,

- **starting to look at an innovation just like at a theorem proving problem is the key to success.**

Thus we would very much like to introduce this new approach as a solid basis for all kinds of investigations in inno-management in general and as

- **a sound basis for a new most promising knowledge management methodology for innovation and R&D-management purposes.**

We are pretty confident, that any reader of this chapter will realize the enormous potentials of a much more rational and profound use of experience and/or knowledge[xxiii] in any R&D-pipeline. As we will be able to show at the end of this chapter (see chapter 4.4 and especially Fig. 39, Fig. 40 and Rule 13),

- **knowledge and its proper management is the single most valuable and most productive resource in any innovation endeavor.**

In the following paragraphs, we will try to derive a sound foundation for an inno-knowledge management, for an appropriate logical and for an empirical calculus to guide and to base suitable IT- and process-support systems upon.

---

[xxiii] Knowledge and knowledge management is according to our definition quite equivalent to what one would have called „experience" of a scientist or of an engineer before.

> **Remember:**
> We are at best amateurs in math, logic and theorem proving. Thus we do not at all claim, that our proposed approach is consistent nor complete in any respect. We rather would like to have the reader interpret our approach as an invitation to specialists and to other people more gifted in these domains than we are.
>
> We would very much like these people to apply and extend their methods from these disciplines to inno-management problems. We do hope, that taking the best of these worlds and sciences should lead to a considerable boost for the innovation management science, the corresponding IT-support systems and for practice in inno-management in the R&D-departments and firms as well.

To start our endeavor, we would like to ask the following question easy to pose but no quite so easy to answer precisely:

> *(Q 12)* **What is the structure and the impact of knowledge in innovation management?**

Let us start to answer this question, by thinking a little about the basic difference, benefit or value knowledge can introduce into inno-management. According to the logic paradigm and the control paradigm of our inno-gem, the only difference, which could be achieved by bringing in some (predictive) knowledge into the innovation process, is a reduction in search costs expressed in time, in efforts, in budget or what so ever. This leads us directly to the following Definition 17 for the value of knowledge $V_K(IP, K)$ for an innovation pipe IP or for an innovation project $Ip_k$ in the next paragraph.

### 4.3.1   The value-of-knowledge concept and its consequences

*Definition 17  -  the value of knowledge $V_K(IP, K_j \neq \varnothing)^{xxiv}$ for an innovation Pipe IP*

(D 17)   $V_K(IP, K_j)$   $=$   $C_F(IP, K_j = \varnothing) - C_F(IP, K_j \neq \varnothing)$
            $=$   $Ci(IP(\lambda(K_j = \varnothing))) - Ci(IP(\lambda(K_j \neq \varnothing)))$

   *with $C_F(IP, K_j = \varnothing)$   $=$   failure costs without knowledge*
   *and $C_F(IP, K_j \neq \varnothing)$   $=$   failure costs with knowledge $K_j$*
   *and $C_i(IP(\lambda(K_j \neq \varnothing))$   $=$   costs of information with knowledge $K_j$*

With some minor modifications we can easily expand this concept $V_K(IP, K_j)$ of the value of knowledge to be applicable not only for inno-pipes IP but as well for innovation projects $Ip_k$ too. We just have to distinguish between 2 cases. For the first case, we conclude from (D 17), that for unsuccessful innovation projects ($Ip_k \in IF$; with IF = set of failing projects):

(D 17a)   $V_K(Ip_k \in IF, K_j)$   $=$   $costs(Ip_k, K_j = \varnothing) - costs(Ip_k, K_j \neq \varnothing)$

For the second case we define that for successful innovation projects ($Ip_k \in$ IS; with IS = set of successful projects):

(D 17b)   $V_K(Ip_k \in IS, K_j)$   $=$   $C_i(Ip_k(\lambda(K_j = \varnothing))) - C_i(Ip_k(\lambda(K_j \neq \varnothing)))$
            $=$   $costs(Ip_k, K_j = \varnothing) - costs(Ip_k, K_j \neq \varnothing)$

The problem with the definition above is our basic a-priori versus a posteriori knowledge problem in innovation management. But statistics and the natural latency of innovation-pipes IP do help us quite a bit in this case. This is due to the fact, that we should see a clear tendency of a dropping $C_F(IP, K_j)$ once we introduce some knowledge $K_j \neq \varnothing$ into our innovation pipe IP.

If we consider single projects only, there are no statistics and on top of that, there is in general no alternative without the application of knowledge ($K_j = \varnothing$) available either. Nevertheless, we do think it is appropriate to use expectancy values of typical failure- and success-costs for certain classes of projects $Ip_k$ as a fair estimate of those costs. By this we at least get an idea, whether or not a knowledge $K_j$ would have been beneficial or not. This again is nothing but applying once more our most fundamental cost of information principle (L 6). Here we do it to judge the value of some knowledge $K_j \neq \varnothing$ for innovation projects and/or processes. This line of thinking does lead us directly to Lemma 14 underneath.

---

[xxiv] From now on we will use the $\varnothing$-character as an abbreviation for the "empty set" {}!

*Lemma 14 - the value $V_K(K_j)$ of any real inno-knowledge $K_j \neq \varnothing$ is always positive.*

(L 14)   The value $V_K$ of any real knowledge $K_j \neq \varnothing$ introduced into an inno-pipe **IP** or to an inno-project **Ip$_k$** must be positive:

$$V_K (IP, K_j) > 0 \quad \text{and/or} \quad V_K (Ip_k, K_j) > 0$$

**Proof of Lemma 14 - the value of innovation knowledge (1):**

From the point of view of any theory of a knowledge management for innovation purposes, this lemma rather has the status of an axiom. It does, for the first time, give a criteria to judge, whether or not some sentence or rule $K_j$ is to be considered to be a (valuable) knowledge at all. Such a clear, operational and testable definition of knowledge had been missing so far in inno-knowledge management, at least to our knowledge. Thus this lemma is a key part, an axiom of the proposed sound foundation of knowledge management. It thus cannot be derived formally within that context.     **Q.E.D.**

**Proof of Lemma 14 - the value of innovation knowledge (2):**

From the point of view of our innovation theory, the Lemma 14 is quite easy to prove. It is just an application of the costs of information principle (L 6) or even more basic of the economic principle to innovation-pipes or -projects. Now let us assume this lemma does not hold, then some knowledge $K_j$ introduced would lead to an augmentation of the necessary investments costs $C_F$ or $C_S$ without affecting the returns/profits Mp of our pipe IP or the respective projects Ip$_k$. The respective RoI will be diminished and we all would be economically better off not having used the supposed knowledge $K_j$ in the first place, even without paying for it. This is most obviously unreasonable and uneconomical and thus the above Lemma 14 really does hold.     **Q.E.D.**

We now do have an appreciation of the value $V_K (K_j)$ of some knowledge $K_j$ but we still do not understand properly how this value $V_K$ is being created. We can get a first idea on how the introduction of knowledge changes the inno-process by looking at Fig. 36. In this picture we use our familiar "most optimal inno-project proof tree" from chapter 1 (see also Fig. 4) to demonstrate how knowledge does change the search for an inno-success of any project Ip$_k$. It is important to keep in mind that now we have to change the search-direction. To prove the lemmas (L 2) and (L 3) we were looking bottom to the top at the proof-tree. Sitting on the (supposed) success-node Ip$_{N1}$ in Fig. 4 we did try to compute the chances to hit that node coming from the top. To evaluate the effects of a knowledge $K_j$ on that very search, we have to redirect the question and our point of view. Now we are asking, looking from top to the bottom,

- **how big are the chances to find a (technology and market) success
  node in the very part of the tree we are investigating right now?**

In order to answer this question in the most general way possible, we modify our "most optimal proof tree" from Fig. 4 a little. We now do assume every node to have q successors with $q \geq 2$. This is more general and by the way more realistic, as can been seen in following figure:

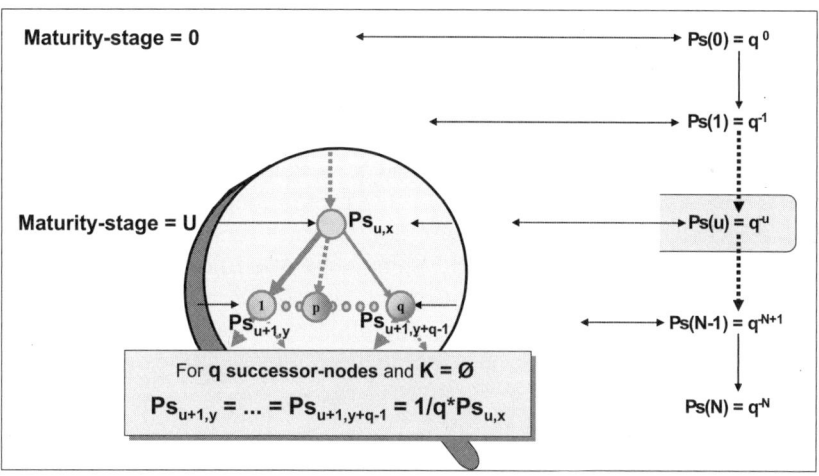

*Fig. 36: The most optimal q-possibilities inno search-tree and the value of knowledge*

Obviously starting from the top node at maturity level 0, everything is possible and thus we get as the corresponding success chance Ps(0)=1. This should be obvious, because we constructed our search-tree such, that it systematically embodies every possible solution of our inno-search $Ip_k$[xxv] and of the corresponding proof problem. Additionally we assume the corresponding proof-tree to be regular (a node is either terminal or it does have q successors), to be of depth N corresponding to the number of necessary maturity steps and to be sorted from left to right with the probably more successful nodes to the left (green nodes in Fig. 36). These assumptions, by the way, do not restrict nor alter the general validity of our investigation. Accepting them we see, that without the application of knowledge the success probability $Ps_{u+1,y}$ of an arbitrary successor node (u+1, y) of node $P_{u,x}$ is

$$Ps_{u+1,y} = 1/q * Ps_{u,x}$$

---

[xxv] Now we benefit from the logic paradigm of the inno-gem by interpreting our inno-project $Ip_k$ as a proposition with its corresponding proof-tree.

Let us call $P_{u,x}$ a father node which has q successor nodes $P_{u+1,y+z}$ with $0 \leq z \leq$ (q-1). In the corresponding tree coordinate system (u, x), u defines the maturity level ($0 \leq u \leq N$) or a search tree depth and x defines the node position in number of nodes at this maturity level (see also Fig. 36) counted from left to right. This and the formula above is consistent with the overall success probability of an N-step and q-successor proof-tree Ps(N) being:

$$Ps(N) = q^{-N}$$

If we now do introduce some knowledge $K_j \neq \emptyset$ at the arbitrary but fixed project-node $Ip_{u,x}$ with the corresponding success-chance $Ps_{u,x}$, we do come to a situation sketched in Fig. 37. But before discussing this picture in more detail, we would like to introduce some definitions first.

On the basis of the inno-gem, we derived in chapter 2 some quite convenient formulas to compute the costs Cs(IP) for the innovations successes (D 10), the costs Ci(IP) of information and the maximum efficiency $E_{IP}(IP)$ achievable for an inno-pipe IP. With some minor modifications we will now derive a calculation scheme/algorithm for the value $V_K$ of some knowledge $K_j$ introduced into an inno-pipe IP or applied for an arbitrary inno-project $Ip_k$. According to (D 10), the costs for the innovation successes are:

$$Cs(IP) = \int\limits_{}^{TtM_{IP}} Ca_{IP}(t) * Ps_{IP}(t)dt = \sum_{i=0}^{N-1} Ca_{IP}(i+1) * Ps_{IP}(i+1)$$

Where $Ca_{IP}(t)$ are the (R&D-) capacity-costs at time t. Transposing (D 10) to any single successful project $Ip_k$ out of which IP is a set, we get for the innovation success costs for this very project $Ip_k$:

$$Cs(Ip_k) = \sum_{i=0}^{N-1} Ca_{Ip}(i+1) * Ps_{Ip}(i+1)$$

If we now assume an (at least per inno-step i to i+1) constant cost- and/or capacity-strategy we get:

$$Cs(Ip_k) = \sum_{i=0}^{N-1} Ca_{Ip}(i+1) * Ps_{Ip}(i+1) = \sum_{i=0}^{N-1} C_{\Delta M}(i) * Ps_{i+1,x}(Ip_k)$$

With the maturity-step costs $C_{\Delta M}(i)$ for the project $Ip_k$ from Definition 18 below and the search-tree success probability $Ps_{u,x}(Ip_k)$ from Definition 19 below, we will now be able to describe the effects of some knowledge $K_j$ introduced into any inno-search as sketched in the Fig. 37 below.

*Definition 18 - the maturity-step costs $C_{\Delta M}(u)$*

*(D 18)* $C_{\Delta M}(u)$ = *costs ($Ip_k$, step ($u$)) - costs ($Ip_k$, step ($u$-1))*
= *costs to do a transition from maturity-level ($u$-1) to the next maturity level ($u$) in an Inno-project $Ip_k$*

*Definition 19 - the search-tree success-probability $Ps_{u,x}$ ($Ip_k$) of project $Ip_k$*

*(D 19)* $Ps_{u,x}(Ip_k)$ = $Ps(Ip_k$, step ($u$))
*at node $x$ of its inno-search tree counted from left to right*

What is missing to fully understand the effect of knowledge in an inno-success search-tree is the exact definition of the "credibility" cred($K_j$) of some knowledge $K_j$ introduced. This is mandatory, cause there is no certainty, no "absolute knowledge" available in the innovation domain at all! The only certainty we ever will reach are logical conclusions. They in turn are either based on more or less certain assumptions or on axioms, which cannot be proved at all. To stay correct, we just have to account for their respective "uncertainty" by introducing the notion of the "credibility" cred($K_j$) of the respective knowledge $K_j$ introduced. Thus we define the credibility cred($K_j$) of some knowledge (proposition, rule, test, condition etc.) at any given application scenario $A_{sc}$ as follows:

*Definition 20 - the credibility **cred($K_j$)** of some knowledge $K_j$*

*(D 20)* **cred($K_j$, $A_{sc}$)** = *Probability that the application of knowledge $K_j$ under the scenario $A_{sc}$ leads to some result $R$. predicted by $K_j$*

Last but not least we should define the "application scenario" $A_{sc}$. We need this notion to describe when which knowledge $K_j$ is applicable in which situation (for which search-node ($u$, $x$)) during the search for an inno-success of the supposed project $Ip_k$. We thus define the $A_{sc}$-condition as follows:

*Definition 21 - the application scenario $A_{sc}$ ($K_j$, $I_{u,x}$, $D_k$)*

*(D 21)* $A_{sc}$ ($K_j$, $I_{u,x}$, $D_k$) = *Set of successor-nodes $\{I_{u+1,y}\}$ of some arbitrary inno-node $I_{u,x}$ within the knowledge-domain $D_k$ where the application of the knowledge $K_j$ does lead to the predicted results $R$.*

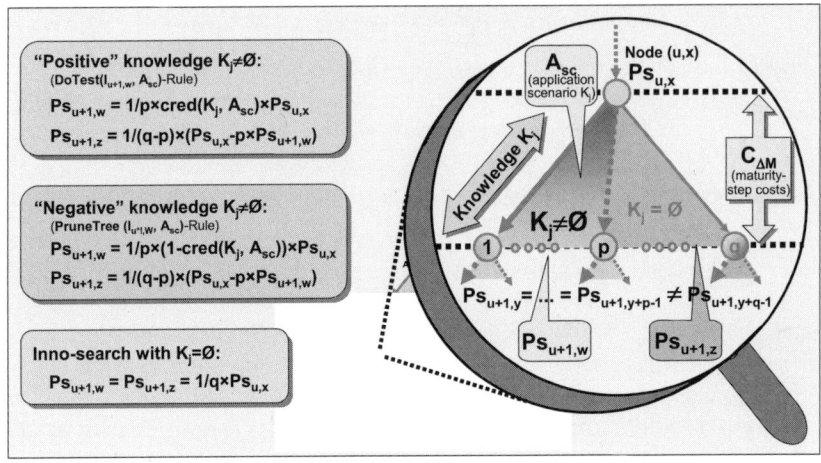

*Fig. 37: The inno-search tree and the effect of an introduced knowledge $K_j$*

### 4.3.2   The two most basic kinds of inno-knowledge

It is a quite worth while endeavor to study Fig. 37 in a very detailed way. This figure does tell us the **4 most essential effects of the introduction of knowledge to any search for an innovation-success:**

1) The **introduction of a knowledge $K_j$** at some inno search tree node $I_{u,x}$ **does separate the set of his successor nodes $I_{u+1,y}$ of into 2 classes,** the "knowledge nodes" $Ps_{u+1,w}$" (green area in Fig. 37), where the application scenario $A_{sc}$ of $K_j$ holds and the other ones (grey area in Fig. 37).

2) The **success probabilities $Ps_{u+1,w}$** of the p "knowledge nodes" are essentially **determined by the credibility cred($K_j$, $A_{sc}$)** of the respective knowledge $K_j$ applicable.

3) The respective **success probabilities $Ps_{u+1,z}$** of the (q-p) successor nodes not directly affected by $K_j$ are **indirectly affected by** the introduction of the very knowledge **$K_j$ via its respective "lack of credibility" (1-cred($K_j$, $A_{sc}$))** which changes their respective success chance $Ps_{u+1,z}$.

4) There **are only two sorts of knowledge $K_j$ ever being applicable to innovation:**

   4a) The **"positive" knowledge DoTest ($I_{u+1,w}$, $A_{sc}$)** which gives the directions to the inno-search and, in case, advises to evaluate a certain set of technology conditions $T_j$ and of corresponding market conditions $M_j$

**4b)** The **"negative" PruneTree ($I_{u+1,w}$, $A_{sc}$) knowledge,** which simply tells us that node (u+1, w) and with it all its successor nodes (u+2, y) fail to meet the necessary inno-success expectations. Thus it is a terminal one and may be discarded from the search.

Especially the last one of those 4 effects of the introduction of knowledge into inno-processes does lead us to the following lemma, which is together with Lemma 14 another part of the proposed sound foundation of inno-knowledge management:

*Lemma 15 - the two kinds of inno-knowledge*

*(L 15)* **There are just 2 kinds of knowledge applications in inno-management. The positive "DoTest ($I_{u+1,w}$, $A_{sc}$)" knowledge, which gives advice which node with which technology and market (pre-) condition to test next and the "PruneTree ($I_{u+1,w}$, $A_{sc}$)" knowledge which gives criteria when to better stop an inno-search in a certain direction.**

---

**Proof of Lemma 15 - the 2 kinds of inno-knowledge**

If one accepts, that any inno-search can be described as a proof-tree search problem too, one immediately realizes that there are just 2 options to proceed in any tree-search problem:

**1)** Stop and prune the corresponding sub-tree, preferably cause you found a terminal node.

**2)** Proceed in a (certain) direction to the corresponding successor nodes (u+1, w).

These only 2 possibilities available for any tree-search problem directly correspond to the respective only 2 options possible for the application of knowledge in innovation management. **Q.E.D.**

---

This lemma does render us the basis for the following definitions of 2 basefunctions or -algorithms which do structure any kind of a knowledge applicable in inno-management. As has been said before, these algorithms are closely linked to the 2 possibilities offered by any tree-search algorithm, do or do not proceed (prune this part of the tree) with your search in that direction to successor node (u+1, w).

*Definition 22 - the basic DoTest ($I_{u+1,w}$, $A_{sc}$) inno-knowledge algorithm*

(D 22) **IF**      *($A_{sc}$ ($K_j$, $I_{u,x}$, $D_k$) $\supseteq$ {$I_{u+1,w}$}) = **TRUE***

        **THEN**      *CONTINUE_SEARCH_WITH_NODE ($I_{u+1,w}$)*
                   *INCREASE_CRED (DoTest ($I_{u+1,w}$, $A_{sc}$))*

        **ELSE**      *SEARCH_OTHER_NODE ($I_{u+1,z}$)*     *with $z \neq w$*

*Definition 23 - the PruneTree ($I_{u+1,w}$, $A_{sc}$) inno-knowledge algorithm*

(D 23) **IF**     *($A_{sc}$ ($K_j$, $I_{u,x}$, $D_k$) $\supseteq$ {$I_{u+1,w}$}) = **TRUE***

      **THEN**   *STOP_SEARCH_WITH_NODE ($I_{u+1,w}$)*
                *INCREASE_CRED (DontProceed ($I_{u+1,w}$, $A_{sc}$))*

      **ELSE**   *SEARCH_OTHER_NODE ($I_{u+1,z}$)*     *with $z \neq w$*

To make life a little more complicated, we would like to mention here a slight augmentation to the DoTest ($I_{u+1,w}$, $A_{sc}$) algorithm. It directly embodies some additional specific knowledge $K_s = \{T_s, M_s\}$ on certain sets of technology $\{T_s\}$ and/or market conditions $\{M_s\}$ to be fulfilled for inno-success in that very application domain $A_{sc}$. Let us call this algorithm DoTestKnowledge ($I_{u+1,w}$, $A_{sc}$, $K_s$). We should be aware, that this algorithm could be replaced by a sequence of appropriately selected DoTest-searches. Thus it is not a real base algorithm, but a comfortable extension to embody new knowledge and to learn from its respective effects to guide the tree-searches.

*Definition 24 - the DoTestKnowledge ($I_{u+1,w}$, $A_{sc}$, $K_s$) augmented inno-knowledge algorithm*

(D 24) **IF** *($A_{sc}$ ($K_j$, $I_{u,x}$, $D_k$) $\supseteq$ {$I_{u+1,w}$}) = **TRUE***

     **THEN**

       **[IF**   *($\forall_s$ ($T_s(I_{u+1,s})$ $\in K_s$ = **TRUE**)*
           **AND**   *$\forall_s$ ($M_s(I_{u+1,s})$ ) $\in K_s$ = **TRUE**)*
           **AND** *{$I_{u+1,s}$} $\subseteq A_{sc}$ ($K_j$, $I_{u,x}$, $D_k$) )*

       **THEN**
          *CONTINUE_SEARCH_WITH_NODE ($I_{u+1,s}$)*
          *INCREASE_CRED (DoTestKnowledge ($I_{u+1,w}$, $A_{sc}$, $K_s$))*

       **ELSE**
          *CONTINUE_SEARCH_WITH_NODE ($I_{u+1,w}$) with $w \neq s$* **]**
          *INCREASE_CRED (DoTest ($I_{u+1,w}$, $A_{sc}$))*

     **ELSE**   *SEARCH_OTHER_NODE ($I_{u+1,z}$)*    *with ($z \neq w$ **AND** $z \neq s$)*

After these lengthy definitions of inno-search or -knowledge algorithms, we would now like to start to discuss a little their respective impacts on the inno-search, on the respective success-chances and on the respective credibility of the knowledge applied. For a much more detailed discussion of these questions, on the setting up and on how to make the respective knowledge work, we again would like to refer to [5], where these question are treated in great detail.

**To Definition 21 - application scenario $A_{sc}$ ($K_j$, $I_{u,x}$, $D_k$) of some knowledge $K_j = \{T_j, M_j\}$:**

The application scenario $A_{sc}$ is the selection condition, which does tell us when to apply some knowledge $K_j$. Thus it is not information to guide the search but to guide the application of knowledge, which is by the way a search problem quite similar to the inno-search. Here again we do have a structured and regular search space (see Fig. 38 and Fig. 41) and a corresponding search problem to find appropriate sets of knowledge $K_j$ applicable at the inno-search node (u, x) under investigation. Looking at the generation and at the augmentation of inno-knowledge, we have to ask ourselves,

- **how do we know that a knowledge (e.g. $A_{sc}$-condition) is a valuable resource or not?**

Referring to our value of knowledge concept (see (D 17) and (L 14)), we can say, that an $A_{sc}$-condition is relevant the more often it triggers the application of some DoTest- or PruneTree-rule. Thus, on top of an estimated start-relevance (e.g. from an expert-circle evaluation as described in [5]), it is a good idea to embody a frequency counter for each $A_{sc}$ into a corresponding KM[xxvi]-support system. This does help to verify the start-relevance and it forms a solid basis for some quite elaborate learning and optimization schemes for the respective KM-systems.

**To Definition 22 - basic DoTest ($I_{u+1,w}$, $A_{sc}$) rule:**

The nice feature about this algorithm or rule for inno-search problems is the fact, that this "positive" knowledge $K_j$ does directly guide the inno-search via improving its respective success probability Ps

$$\text{Ps}_{u+1,w} = 1/p * \text{cred}(K_j, A_{sc}) * \text{Ps}_{u,x} > 1/q * \text{Ps}_{u,x}$$

of the respective p selected possible successor nodes (u+1,w) of father node (u, x). With some minor calculations we see, that, to be valuable, the credibility of $K_j$ just has to be

---

[xxvi] From now on we will use **KM** as an abbreviation for **K**nowledge **M**anagement

$$\mathbf{cred(K_j, A_{sc})} \;\; > \;\; \mathbf{p/q}$$

with   p = number of successor nodes (u+1, w) selected by $K_j$ and
     q = number of possible successor nodes (u+1, z) of father
     node (u, x).

Again we see, that good common thinking and our theory does render the
same results. In general a valuable inno-knowledge $K_j$ should have a credibility $cred(K_j, A_{sc}) > 50\%$ and it should be rather selective (p≤q) too. But
supposed we do have some very specific knowledge $K_j$ with p<<q, we could
also work quite well with a much more insecure ($cred(K_j, A_{sc})$<<50%)
knowledge $K_j$. Thus it obviously pays to look into the logics and maths of
inno-knowledge! A fair rule of thumb criterion for positive inno-knowledge
$K_j$ is to go for credibilities in the 2:1 to 4:1 region corresponding to

$$\mathbf{70\% \; \le \; cred(K_j, A_{sc}) \; \le \; 80\%}$$

As far as the credibility values for the DoTest-, DoTest knowledge and the
Prune-Tree- algorithms are concerned, we do have a validity problem, just
like with our $A_{sc}$ condition before. Thus we strongly propose any theoretical
as well as real world KM-System or concept to

- **provide an elaborate credibility-evaluation scheme, e.g. by embody-
  ing hit-frequency counters to account for the relative frequency of
  true/verified applications of $K_j$ with respect to all real or possible
  application trials.**

The function **INCREASE_CRED (function, arg1, arg2)** in the respective
algorithms is just meant as a reminder to perform this task.

### To Definition 23 –   the PruneTree ($I_{u+1,w}$, $A_{sc}$) inno-knowledge algo-rithm:

If one does not have a positive advice (DoTest-Rule) what to do next in an
inno-search, there still may be some "negative" knowledge on what to avoid
in any case. We are personally rather tempted to state, that there always must
be some "negative" knowledge available, cause

- **avoiding to take irresponsible risks is a precondition for maximizing
  profits in any real world economic endeavor.**

This is just the same as any quality gate system described in chapter 3.3 (see
especially Rule 6) does perform. Thus we can say, that

- **any inno-quality gate System $Qg_k$ is by nature an implementation of
  a KM-System comprised of only a set of  PruneTree ($I_{u+1,w}$, $A_{sc}$)
  rules.**

This line of thinking does comply completely with our introduction of a Qg-System as a best (lower boundary) estimate for the unknown success function $Ps(Ip_k)$ or $Ps(IP)$ of an inno-project $Ip_k$ or an inno-pipe IP. On top of that, we can derive some quite interesting criteria for the values of the credibility $cred(I_{u+1,w}, A_{sc})$, the applicability $A_{sc}(K_j, I_{u,x}, D_k)$ and the "subsumption" logic for quality-gate and/or PruneTree-systems or algorithms:

1) The credibility of a PruneTree$(I_{u+1,w}, A_{sc})$ rule must be quite high, cause they cut-off the corresponding sub-trees and their respective inno-success options from any inno-search. A good rule of thumb for an acceptable credibility is a 9:1 to a 19:1 hit rate corresponding to

$$90\% \leq cred(PruneTree\ (I_{u+1,w}, A_{sc})) \leq 95\%$$

2) The application scenarios $A_{sc}$ of subsequent quality-gates and/or PruneTree applications $Qg_k$ and $Qg_{k+1}$ must be subsuming. This is a logical consequence of following lemma about the respective application scenarios $A_{sc}(K_j, I_{u,x}, D_k)$:

*Lemma 16 - the subsumption rule for application scenarios $A_{sc}$ in any inno-search path*

(L 16)   $$A_{sc}(K_j, I_{u,x}, D_k) \supseteq A_{sc}(K_l, I_{n+1,w}, D_k)$$
with   $K_j$ and $K_l$   = *some negative inno-knowledge*
      $I_{u,x}$   = *father node of $I_{u+1,w}$*
      $D_k$   = *knowledge domain with $D_k \not\subset D_m$ if $k \neq m$*

---

**Proof of Lemma 16 - the subsumption rule for subsequent $A_{sc}$:**
   In any proof-tree search the verification conditions $V_c(u,x)$ and $V_c(u+1,y)$ of subsequent nodes (u,x) and (u+1,y) are conjunctions, they are "and-connected". Thus we can only conclude from the more general $V_c(u,x)$ to the more specific condition $V_c(u+1,y)$ with $V_c(u,x) \supseteq V_c(u+1,y)$. Having introduced and defined our inno-search process as a proof-tree search of $Ip_k$ for the corresponding inno-success proposition $P_s(Ip_k) = 1$, Lemma 16 necessarily does hold.                                    **Q.E.D.**

---

Lemma 16 is enormously important for KM-systems, cause it forms the basis of a general structure of any KM search-space (see also (L 17) and Fig. 38). Just looking at Fig. 37 and at the $Ps_{u+1,w}$-values computed there for "positive" DoTest-rules and "negative" PruneTree-rules, we might already imagine the basic structure of a KM search-space for knowledge applications quite well. This basic structure is very simple and it is consistent with the "monotony condition" for Qg-systems (see Rule 6, (D 6), (L 5), (L 13)) already derived in chapter 1 and chapter 3 on a completely different basis.

This basic structure of an appropriate general purpose inno-KM space can be described as follows:

- **Going top to bottom, from more general to more specific T- and M-conditions (subsuming one another) along any one search path.**

- **Going right to left from the very specific "killer-rules" (the Prune-Tree-rules or quality-gates) to the more and more specific "Do_look_for_these_T&M_conditions_rules" (DoTest- rules).**

This line of thinking leads us directly to Fig. 38 (structure of the KM-search space) and to the basic KM-search space axiom Lemma 17. But before discussing these issues, we would like to spend some more thoughts on the Definition 24 (DoTestKnowledge-rule) first.

### To Definition 24 - the DoTestKnowledge ($I_{u+1,w}$, $A_{sc}$, $K_s$) algorithm:

As we mentioned already, this rule is a slight augmentation to the DoTest Rule (D 22). We did introduce it and we do propose it, to better be able to structure and explain the generation and the verification of new knowledge into any KM-system or search-space or -domain $D_i$.

The new thing in this definition in comparison to (D 22) is the new additional knowledge $K_s = \{T_S, M_s\}$ to be tested or verified. Now we are in KM-paradise, cause we do have some more specific technical and market knowledge or advice to give:

- A set of technology success conditions $T_s = \{T_1,......T_s\}$ and

- A set of market success conditions $M_s = \{M_1....M_s\}$ to be fulfilled.

These conditions $K_s = \{T_s, M_s\}$ have to be met by a search node $I_{u+1,w}$ and, as a logical consequence, by all its successor nodes $I_{u+k,w}$ (with $1 \leq k \leq (N-u)$) too, once we continue the search for inno-success with this node. Following this most general search logic we simply do

- **conclude and/or search from the general to the more specific success conditions $K_s = \{T_s, M_s\}$.**

We see that any inno-success search for any inno-project $Ip_k$ and for any inno-knowledge domain $D_i = \{K, I_{11}, I_{NM}\}$ essentially does have a quite comparable structure and does have quite comparable properties as sketched in Fig. 38 and in Fig. 41. This leads us directly to Lemma 17 about the regular structure of any inno-knowledge space in the following paragraph. As a personal comment to this lemma, we would like to state, that we would rather consider it to be at least a (most fundamental) axiom of any theory about inno-knowledge management (see also Fig. 38).

### 4.3.3   The proposed most basic structure for an inno-knowledge space

*Lemma 17 - the regular structure of any inno-search or -knowledge space*

*(L 17)   A search- and/or knowledge-space $D_i$ for any inno-project $Ip_k$ can be organized such that*

- *any inno-success condition and/or knowledge $\boldsymbol{K_s} = \{T_s, M_s\}$ at a search node $\boldsymbol{I_{u,x}}$ subsumes the success conditions of its $\boldsymbol{q}$ daughter nodes $\boldsymbol{I_{u+k,y}}$ along any search path $\boldsymbol{Sp}$*

$$\boldsymbol{K_s(I_{u,x})} \supseteq \boldsymbol{K_s(I_{u+k,y})} \text{ for any } \boldsymbol{k} \quad \text{with} \quad 1 \leq k \leq (N-u)$$
$$\text{if } \boldsymbol{I_{u+k,y}} \in \boldsymbol{Sp(I_{u,x})}$$
$$\text{with } \boldsymbol{Sp(I_{u,x})} = \text{set of subsequent successor nodes } \{\boldsymbol{I_{u+1,y}},$$
$$\boldsymbol{I_{u+2,z}} ..., \boldsymbol{I_{N,v}}\} \text{ of node } \boldsymbol{I_{u,x}}.$$

- *at least within one set of $\boldsymbol{q}$ daughter nodes $\boldsymbol{I_{u+1,w}}$ of $\boldsymbol{I_{u,x}}$ the coresponding Ps-values can be ordered*

$$\boldsymbol{Ps(I_{u+1,y})} \geq \boldsymbol{Ps(I_{u+1,z})} \quad \text{if } y < z$$

This lemma forms a most solid basis to structure and to organize any inno-knowledge domain $D_i$ as sketched in Fig. 38. Additionally it serves as a scheme for how to generate, test and validate the introduction of any new knowledge $K_j$ into an inno-knowledge domain $D_i$. We thus may conclude, that

- **any inno-knowledge domain $D_i$ can and must be built incrementally.**

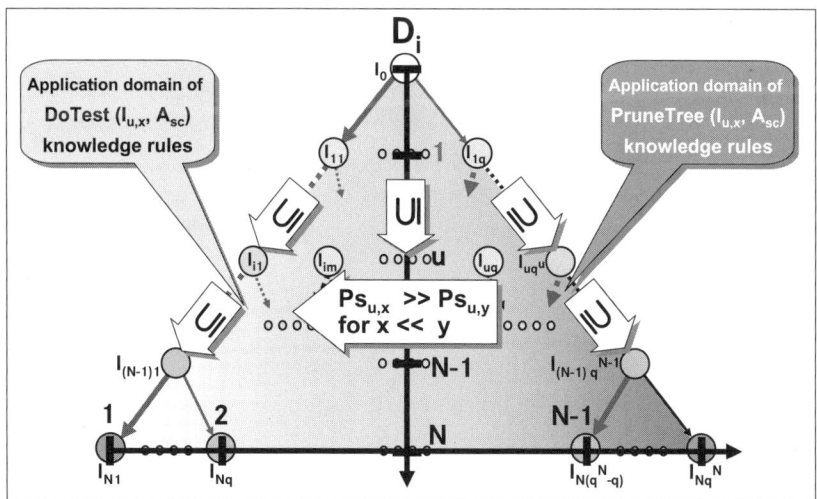

*Fig. 38: The regular structure of inno-knowledge domain $D_i$ and the effects of knowledge $K_j$*

**Proof of Lemma 17 - the regular structure of an inno-search and/or knowledge-space:**

From the point of view of a KM-theory, we would rather consider (L 17) to be an axiom which needs not to be proved and which at least we personally do consider to be quite evident.                                    **Q.E.D.**

Looking on (L 17) from the point of view of our inno-gem, we think its validity is obvious. It just states that one can and that one has to search and to conclude from the more general to the more specific inno-success condition $K_s = \{T_s, M_s\}$. This is even dictated by logic, cause along any one inno-search path Sp all the success-conditions $K_1$ to $K_N$ do form a conjunction with $T_s$ & $M_s = K_1 \& K_2 \&, ..., \& K_N$. Additionally we would like to remark that (L 17) is the equivalent in the logic domain to the monotonic increase conditions (L 4) and (L 5) for the success-function $P_s(Ip_k)$ for any optimally conducted inno-search $Ip_k$ and/or inno-pipe IP.                    **Q.E.D.**

This most well structured inno-knowledge space $D_i$ does render quite some most useful rules and indicators on how to construct, to augment and to run an innovation KM-system. It may sound astonishing, but we are quite sure, that any careful reader of our inno-gem (chapters 1, 3.3, 4.1 and 5.6) would rather expect these 5 key properties of any inno-knowledge space $D_i$ listed below:

1) Any quality gate system Qg(IP) already is a KM-system application for that very inno-pipe.

2) Any quality gate system Qg(IP) has to comply with the structure (e.g. the subsumption-rule) and the rules (PruneTree, ....) for the KM-systems described in this paragraph to work optimally.

3) The next step to enhance the performance of an inno-pipe IP is the introduction of a "positive" knowledge KM-system on top of an existing Qg-system for that very inno-pipe IP.

4) The exploitation of experience (T- and M-knowledge) is the single most valuable asset in any inno-pipe, R&D-department or firm (see also Fig. 39 and Fig. 40).

5) It is irrelevant for the exploitation of knowledge, whether or not it is encoded in "technical" KM-systems (e.g. based on E-BoKs[xxvii] etc.) or in "biological KM-systems", which one normally would call senior-experts, chief-engineers, chief-scientists, etc. It is the consequent and well struc-

---

[xxvii] **E-BoK** = **E**ngineering **B**ook **o**f **K**nowledge is the Chrysler, now the DaimlerChrysler name for a structured data base for sets of engineering knowledge data collections. This and the installed "Tech-Club system" was the reason that Chrysler got in 1999 the highest US award for the implementation of a successful knowledge management system.

tured use (see Fig. 37 and Fig. 38) of this knowledge that counts and that pays (see Fig. 39, Fig. 40 and [5] for additional readings).

To finish this paragraph, we would like to do some test to verify the validity of our KM-approach and its two basic rules DoTest and PruneTree. To do this we collect and discuss their respective results for all possible boundary values of their respective parameters

- **Credibility cred($K_i$, $A_{sc}$)** from 0 (totally insecure) to 1 (sure)

- **Selectivity** (No. of selected daughter nodes) from p=1 (exact) to p=q (general)

in the following table underneath:

| | | Cred($K_j$,$A_{sc}$) = 0 | | Cred($K_j$,$A_{sc}$) = 1 | |
| | | P = 1 | P = q | P = 1 | P = q |
|---|---|---|---|---|---|
| **DoTest** | $Ps_{u+1,w}$ (w= selected node) | **0** | **0** | $Ps_{u,x}$ | $1/q^{*} Ps_{u,x}$ |
| | $Ps_{u+1,z}$ (z= unselected node) | $1/(q-1)^{*} Ps_{u,x}$ | $\infty$ (prune tree) | **0** | **0** |
| **PruneTree** | $Ps_{u+1,w}$ (w= selected node) | $Ps_{u,x}$ | $1/q^{*} Ps_{u,x}$ | **0** | **0** |
| | $Ps_{u+1,z}$ (z= unselected node) | **0** | **0** | $1/(q-1)^{*} Ps_{u,x}$ | $\infty$ (prune tree) |

*Table 2: Ps-values for some boundary credibility and selectivity values*

Obviously our Ps-values and calculations for the boundary values in Table 2 do show quite reasonable results:

- For totally insecure knowledge (cred=0), the selected DoTest and the unselected PruneTree nodes (u+1,w) and (u+1,z) show zero Ps-values. The other nodes are unaffected and have Ps-values as if there had not been any knowledge applied. This is obviously sound

- For guaranteed knowledge (cred=1), the application of DoTest is equivalent to chopping off (pruning) all other nodes and the application of PruneTree does render the expected results. This is for p=q a termination of the search ($Ps_{u+1,z} = \infty$) at node (u,x).

To finalize this chapter we would like to compile our results on the structure and impact of KM for innovation management with the following rules to design a decent inno-KM system.

---

*Rule 12 - the 5 most basic design principles for an inno-knowledge management system*

**R12.1)** The **value $V_k$(IP, $K_j$) of any inno-knowledge $K_j$** element to an inno-pipe IP **always is positive** and **does give orientation** on whether to proceed or not and, in case, into which direction from an arbitrary inno-search node $I_{u,x}$ (see (L 15) and Fig. 36)

**R12.2)** There are just **two basic kinds of inno-knowledge $K_j$**. One is the **"positive" DoTest-** or **DoTestKnowledge-rule** (see (D 22) and (D 24)) rendering information on **which technology and/or market condition to investigate next**, while the **"negative" PruneTree-rule** (see (D 23)) does have to give precise information on **when the respective inno-search better is to be stopped** immediately.

**R12.3)** The **PruneTree-rules do always subsume any corresponding inno-quality gate system QgS.** Thus the respective subsumption, optimization and structure rules for the knowledge space do hold for QgS too.

**R12.4)** Any **inno-knowledge $K_j$ at a node (u, x)** of maturity level u **does always subsume** the rules and/or **inno-knowledge of its respective daughter nodes (u+k, y)** with k≥1 along any one search path.

**R12.5)** At least **within one set of daughter nodes (u+1,y) of any inno-search node $I_{u,x}$** the corresponding **credibility and/or success probabilities $Ps(I_{u+1,y})$ of any inno-knowledge $K_j$ of its daughter nodes can be ordered** left to right **according to** their respective **magnitudes.**

---

## 4.4   Costs & benefits of knowledge – the economic value of experience

In the previous chapter 4.3 we learned about the introduction of knowledge into inno-management, whether it is positive (DoTest-rules) or negative (PruneTree- or Q-gates rules). Now we would like to give the interested reader at least an idea, on how an answer to the following question, most important to inno-economics, could look like:

> *(Q 13)*   **What are the minimal and the potential economical benefits of some knowledge $K_j$ introduced into inno-management?**

Keeping Fig. 38 in mind and just taking a glance at Fig. 39 we see the effects of some knowledge $K_j$ introduced in the inno-search of the assumed inno-project $Ip_k$. It does change an inno-search space $D_i$ considerably. With the introduced knowledge, the search space can be structured in way, that the following properties are obtained:

1) The more likely solutions are accumulated on the left and the less likely are concentrated on the right side (see Fig. 38 and Fig. 39) of the search-space or -tree.

2) The Ps-values at each maturity level k do differ considerably, e.g. from $0{,}8^k \geq Ps_k \geq 0{,}2^k$, which is just the numerical expression for the orientation knowledge being generated.

3) The number of nodes (u, x) and of the respective search-paths $Sp_k$, with a better than 1:1 success chance corresponding to values of $Ps_k \geq 2^{-k}$, are exponentially reduced

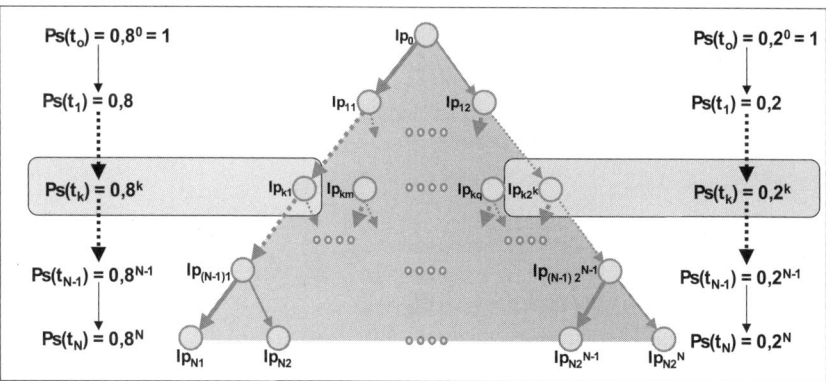

*Fig. 39: Example 80% inno-search knowledge tree and the corresponding costs/probabilities*

For an inno-project, this effect is pretty obvious and a logical consequence of our inno-gem and the corresponding knowledge definitions (D 22), (D 23) and (D 24). To be able to answer question (Q 13) for a complete inno-pipe, we have to keep in mind, that

1) any inno-pipe does separate at any maturity stage u (Q-gate u) an incoming stream of projects into 2 sets IF (failing projects) and IS (successful projects), where only supposedly successful projects from set IS={Ip$_1$, Ip$_2$....Ip$_m$} are evaluated further (see chapter 1.3, (L 6), Fig. 5).

2) this separation is done at any maturity stage u ($\Leftrightarrow$ Q-gate u) on the basis of the corresponding Ps-values Ps(Ip$_k$(u)) $\geq$ Ps$_{min}$(u) being greater than some threshold value Ps$_{min}$(u) for that stage.

3) without any prior knowledge the optimal threshold value for project selection is Ps$_{min}$(u) $\geq 2^{-u}$.

4) there are multiple ways for any project Ip$_k$ at any maturity stage u to achieve a certain value for Ps(Ip$_k$(u)) = Ps$_k$(u), depending on the sequence of its respective positive or negative T$_s$- or M$_s$-evaluations.

Respecting these 4 conditions and doing some combinatorial evaluations on optimal 2-successor node (better/worse) tree-searches, we end-up with a well structured knowledge-application and -benefit scenario, shown in Fig. 40 . In this inno-pipe with an assumed maximum depth of N equidistant maturity stages, we can now discuss how and how much an introduced knowledge does reduce the necessary inno-search effort. In order to be at least as good as without any knowledge K applied, we demand the success probability Ps(u) to be Ps(u) $\geq 2^{-u}$. Together with the combinatorial rules for the success-probabilities we get the following success-condition for an arbitrary search node (u, j):

$$Ps(I_{u,j}) = c^{N-j+1} * (1-c)^{j-1} \geq Ps_{min}(u) = 2^{-u}$$

with   c = constant = cred(K, A$_{sc}$) at the maturity level 1 to N
       u = maturity level ($\Leftrightarrow$ vertical coordinate)
       j = case cardinality, ordered from left (best) to right (worst)
            ($\Leftrightarrow$ horizontal coordinate)

Due to the above, we can calculate the maximum cardinality of the search node (u, j) to be evaluated:

$$j < 1 + N * \ln(2c) / [\ln(c) - \ln(1-c)]$$

With these cardinality numbers we can then calculate the necessary search paths to be evaluated together with their corresponding search-efforts as shown in the following figure (Fig. 40) underneath.

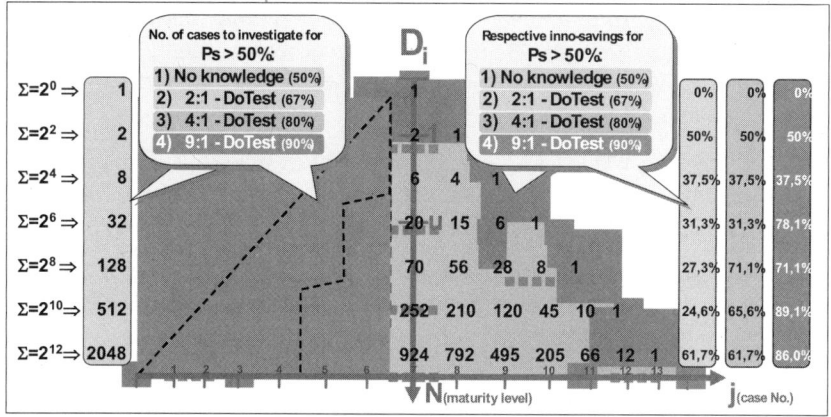

*Fig. 40: The potential inno-search savings of (DoTest-) knowledge in an inno-pipe of depth N*

This modified and necessarily knowledge-based search spaces (green area in Fig. 40) does, as expected, accumulate to the left of the overall search space. It really does cover only a small fraction of the overall search space to be evaluated without the application of knowledge.

All the possible savings listed in Fig. 40 are calculated under the assumption of constant maturity step costs $C_{\Delta M}(u)$ from level 1 to level N all over the whole inno-pipe IP under investigation. With these assumptions, the savings calculated in Fig. 40 are an absolute lower limit in comparison to a "no knowledge applied" (the 50:50 inno-game) tree-search case. There are remarkable jumps/steps in the achievable savings oscillating e.g. from 27% to 62% for a 2:1 DoTest-knowledge application. The explanation for this astonishing behavior is the influence of

- the digitization of the selection process via the N maturity stages (Q-gates)

- the influence of the combinatorial possibilities to reach the same decision node (u, x) via different $T_s(u, z)$ and $M_s(u, z)$ evaluations.

If we prolong that tree to a very large depth N (e.g. N >> 12) we see, that the achievable savings harmonize and home in to the expected values, which are proportional to the credibilities assumed. This observation is in line with some theoretical argument stating, that

- **the maximum efficiency achievable in any "infinite" inno-pipe is proportional to the quality (credibility) of the knowledge applied inside the pipe.**

We refrain from writing the above statement as a lemma, but it has for sure that character and that status. It can be proved easily, going through the combinatorial evaluation leading to Fig. 40. Additionally we see in Fig. 9 on page 36, calculated on a completely different basis, that

- any "no-knowledge" and very long (infinite) inno-pipe cannot have a better efficiency than 50%, which corresponds to its overall statistical 50:50-chance to achieve an inno-success.

On top of that, the same line of thinking does lead us for the case of absolute certainty to the most expected result which is in-line with good common thinking, for we know that

- "god's own inno-pipe" just must have an efficiency of 100%, cause he never does make any mistake.

To finalize, we would like to compile the findings of this paragraph in the following Rule 13 on the value of knowledge for innovation management:

---

*Rule 13  -  4 rules to determine the value and the application of knowledge in inno-management*

**R13.1)** **Knowledge**, whether "positive" ((D 22 or (D 24)) or "negative" (D 23), is the by far **most valuable resource** in inno-management. It is determining the ultimate achievable terminal inno-pipe efficiency (see Fig. 40).

**R13.2)** The knowledge application inside an inno-pipe can be
- **"biological"**, e.g. as chief- or senior-scientists or engineers
or it can be
- **"formal / systematical"** with the support of some appropriate KM-systems
with quite comparable effects but with most different efforts necessary too install and to manage it.

**R13.3)** The maximum **benefit** of some knowledge $K_j$ applied in an inno-pipe is **proportional** to its respective credibility **cred($K_j$, $A_{sc}$)**.

**R13.4)** **Reasonable credibilities** for **"positive"** (DoTest-) knowledge does range from **67% to 80%** (2:1 to 4:1) and for **"negative"** knowledge (PruneTree) from **90% to 95%** (9:1 to 19:1), due to their "pruning" the search space property, which might endanger overall inno-success.

## 4.5   IT infrastructure for a sophisticated inno-knowledge management

In the chapters 4.3 and 4.4 we did derive, apply and exploit for inno-knowledge applications the tree-structure principle for any inno-search problem. Lemma 17 and Fig. 38 did show us in detail, that a "knowledge tree" or "knowledge domain" derived using these properties, is even most regular in size and structure. On the basis of this experience, we may now ask ourselves the following question:

> *(Q 14)   What is a suitable structure for an inno-knowledge space?*

To answer this question, we would again like to exploit the properties mentioned above. We will derive and structure some sort of an ordered "inno-knowledge space" $Ks_I$ together with some appropriate "inno-knowledge coordinate system" for $Ks_I$. Then we will show, that we are much better able to structure and to organize the storage and the retrieval of some inno-knowledge $K_j = \{T_j, M_j\}$ being used in such a structured inno-knowledge space $Ks_I$. Following this line of thinking and focusing on the subsumption rule valid in any inno-knowledge domain $D_i$ from top to bottom, we can now define a suitable inno-knowledge space $Ks_I$ as follows (see also Fig. 41):

*Definition 25  -  the innovation-knowledge space $Ks_I$*

*(D 25)   Any linear structured set of innovation-knowledge domains $D_i$ according to Lemma 17, which does not subsume one another, does form an appropriate **inno-knowledge space $Ks_I$***

$$Ks_I = \{D_1, ..., D_i, ..., D_m\}$$
*with*    $1 \leq i \leq m$    *and*    $D_i \not\subset D_k$ *if* $i \neq k$

*$Ks_I$ does have the cardinality $m$ and it has 3 inno-knowledge coordinates for any innovation-knowledge $K_j$ within $Ks_I$.*

$$K_j => Ks_I(x, y, z) = (D_i, M_u, N_k)$$
*with*   $D_i$   =   *number of innovation-knowledge domain $D_i$*
     $M_u$   =   *maturity level of $K_j$ within $D_i$*
     $N_k$   =   *node-number $N_k$ of $K_j$ at maturity level $M_u$ counted from left to right*

With this definition of an innovation-knowledge space $Ks_I$ we do have a most comfortably structured storage and retrieval (coordinate-) system for inno-knowledge and for KM-systems as well, as shown in Fig. 41. Let us now suppose we organize/store any (new) knowledge $K_j(D_i, M_u, N_k)$ according to the ranking of its respective Ps-value or credibility $\text{cred}(K_j)$ e.g. according to the relation

$$\mathbf{cred}(K_j(D_i, M_u, N_k)) \geq \mathbf{cred}(K_j(D_i, M_u, N_l)) \qquad \text{if } N_k > N_l$$

Respecting this relation, we automatically prevail in any inno-knowledge domain $D_i$ and thus in our inno-knowledge space $Ks_I$ too the total ordering of all the knowledge $K_j$ stored. With these prerequisites, it is a quite easy task to construct and implement a knowledge storage and retrieval system of at least a decent performance. This is illustrated in Fig. 41 below too.

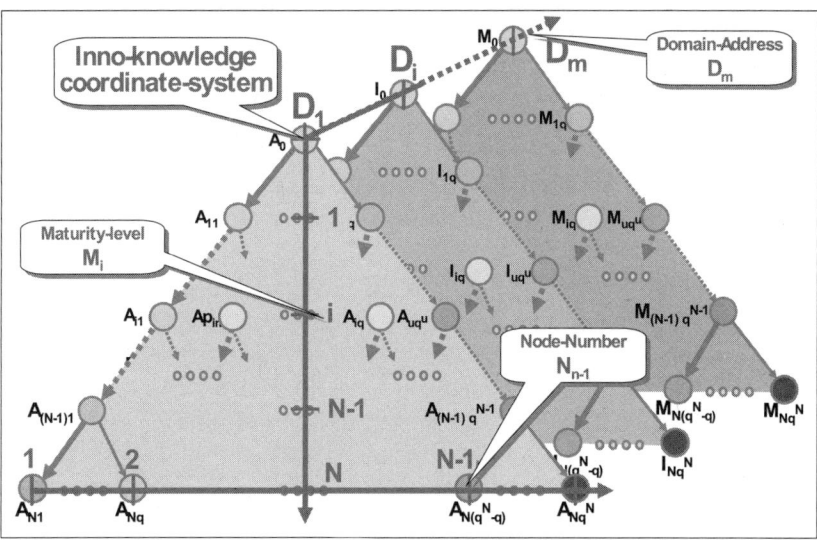

*Fig. 41: The proposed inno knowledge-space coordinate system*

Such a regular and ordered knowledge space makes life real easy for KM-systems to search and retrieve the knowledge $K_j$ most relevant to a specific inno-search problem. Once this knowledge $K_j\{T_j, M_j\}$ is applied at the assumed search node $I_{u,x}$, it can give there quite specific technical $(T_j)$ and market $(M_j)$ orientation and thus help to increase the success-chances $Ps(I_{u,x})$ substantially. Looking a little deeper into the mechanism and the mechanics of knowledge generation, storage and retrieval (see [5]), we have to admit that:

- **the production mechanisms and, in most cases, the quality and the structure of technical $\{T_j\}$ and market $\{M_j\}$ knowledge is quite different.**

This could cause some most serious problems to our KM-storage and re-trieval mechanism (see (D 25) and Fig. 41), once we are not able to over-come the respective differences in structure and quality of these types of inno-knowledge. Knowledge-quality is in general corresponding to some depth/validity in our inno-search trees. This does allow us, just by introduc-ing virtual technology ($T_{j,k}$) and/or market ($M_{j,k}$) success-condition nodes, to always be able to overcome this problem mentioned above. As being illus-trated in Fig. 42, the introduction of the concept of an

- **innovation hyper-node $I_{j,k}$** which at any search-tree or knowledge-domain node (j, k) does represent the **conjunction of the respective technology ($T_{j,k}$) and market ($M_{j,k}$) success conditions**,

does allow us to most effectively overcome the structural "incompatibility" of technical and market knowledge. Using this concept, we have a **unified approach to store, retrieve and compile** these in general **most different kinds and sources of relevant inno-knowledge.** Due to the mentioned properties of inno-search trees, we can introduce hyper-nodes at any time and/or situation. We can do that, via the introduction of intermediate (vir-tual) technology ($T_{j,k}$) and/or market ($M_{j,k}$) nodes, which by definition do not alter the respective Ps-values $Ps(T_{j,k}\&M_{j,k})$ at all. The (virtual) duplication of already verified knowledge/conditions $\{T_{j,k}, M_{j,k}\}$ is just one, most easy way to do so (see Fig. 42).

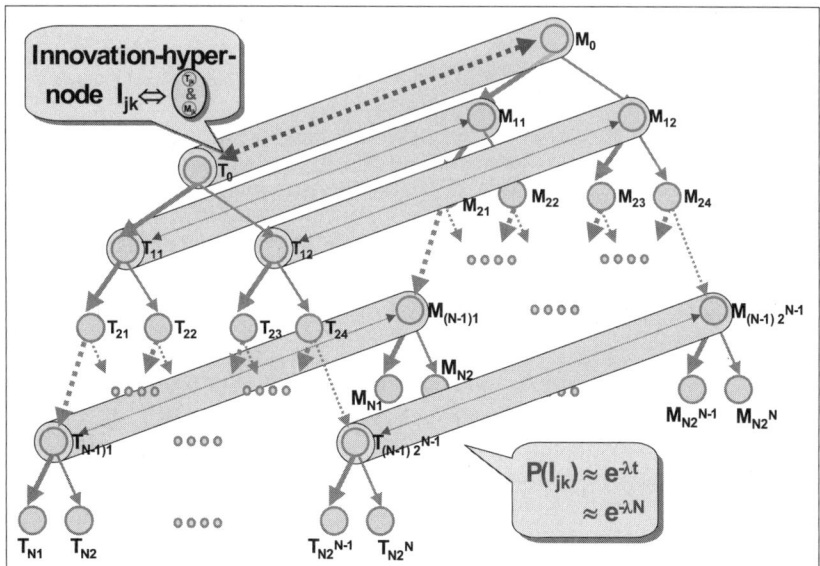

*Fig. 42: The innovation hypernode model for the inno knowledge-space management*

With these elements one should always be able to construct at least a decent KM-support system for virtually any set of inno-projects and/or inno-pipes. What is common to all these KM-systems, starting from the most simple and basic PruneTree- or Q-gate systems up to the most elaborate ones one can ever imagine, is the fact, that

- **any KM-support system inherently must be based on project data.**

Thus it is based on the existence of project management support systems in the envisaged inno-pipe and one will run into serious problems once such systems are not already installed in that inno-pipe. We really do want to emphasize this point cause,

- **without project success-data and success-histories, there is no decent KM-support possible.**

How an elaborate project management support system (PM support-system) could look like is shown in Fig. 43. This figure shows the hierarchical PM- and PM-support IT-systems of DaimlerChrysler corporate research. This PM-system architecture is one of the most comprehensive, most elaborate and most structured systems we do know in industry. At least we do consider this system architecture to be quite a solid basis to do the next steps necessary and possible in inno-management.

To summarize our investigation on IT-system to support inno-management and inno-KM application we would like to give the following rules:

---

*Rule 14 - the 3 most basic design rules for inno-KM support systems*

**R14.1)** Any efficient end effective inno-KM support system has to be based on an (existing) inno-project management and on a corresponding IT-support environment.(see e.g. Fig. 43)

**R14.2)** Any set of well structured and ordered inno-knowledge items $K_j$ can be organized into a regular and well ordered inno-knowledge space $Ks_l(D_j, M_u, N_k)$ with the three coordinates inno-Domain $D_j$, maturity level $M_u$ and node number $N_k$ (see (D 25) and Fig. 41).

**R14.3)** Any technology $(T_{j,k})$ and market $(M_{j,k})$ knowledge space and/or domain can be fused to an integrated inno-knowledge domain $\mathbf{D_j}$ by introducing the concept of an inno-hypernode or-tree $I_{j,k}$ according to Fig. 42.

---

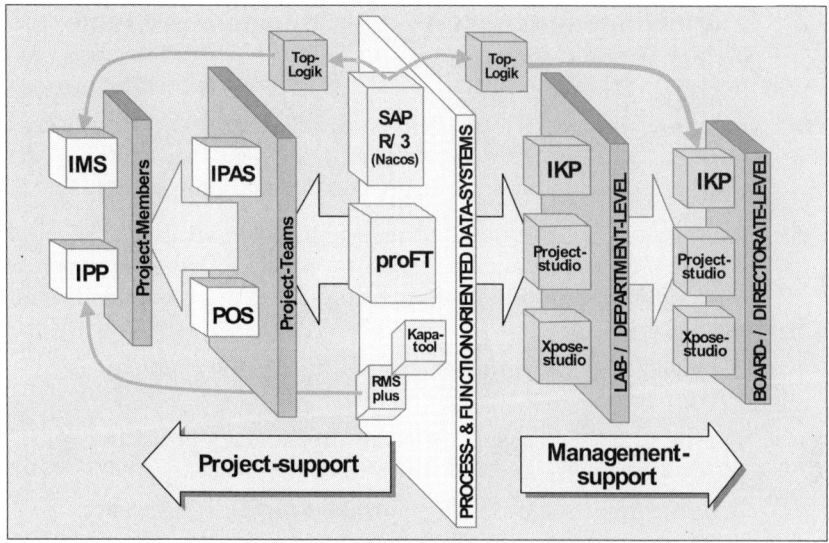

*Fig. 43: An example project management support system architecture (DC R&T)[xxviii]*

[xxviii] Source: Proceedings of the IPMA conference 05. - 06.06.2002 in Berlin

## 4.6 Intermediate summary 4 – the inno-gem and inno-project management

**S4a)** There is a blueprint for any inno-pipe and/or -project, the **inno-maq** (see Fig. 31), describing the **kind and the sequence of the just 6 most basic steps to take** in order to come **to an inno-success.**

**S4b)** There are **only 3 basic inno-strategies possible** (see also (L 7)). Each of which proceeds through the inno-maq **with a different sequence of steps and with a quite different focus.** These most basic innovation strategies are

- **T-push**      focusing on its **80% marketing problem** (see (L 7), Rule 9).

- **D-pull**      focusing on its **80% technology problem** (see (L 7), Rule 10).

- **incremental**   showing large **product-/process-differences** (see (L 7), Rule 11).

**S4c)** On the **basis of the inno-gem** and the statistical filter approach proposed **one can derive sound principles, axioms, rules and even data structures suited for a** substantially **improved inno-knowledge management** (see chapters 4.3 to 4.5).

**S4d)** **Any knowledge $K_j$ introduced into any inno-search or -pipe IP must have a positive "orientation value" $V_k(IP, K_j)$** (see (L 14), Fig. 37 and Rule 12).

**S4e)** There are **only 2 kinds of inno-knowledge applications possible** (see Fig. 37) for any inno-pipe. These are the

**"positive" DoTest-rules**    ⇔ **giving a (T&M-) search directions** (see (D 22) and (D 24)).

**"negative" PruneTree-rules** ⇔ **telling when to better stop** a search in a certain direction see (D 23)).

**S4f)** **All relevant inno-knowledge $K_j$ can be tree-structured and it has to fulfill the subsumption rule** along any one search path, always concluding from the more general to the more specific inno-success conditions (see (L 16) and (L 17)).

**S4g)** Any **knowledge $K_j$ introduced** into an inno-search or pipe **"tilts" the search space** such that one gets a **gradient in** the respective success-probabilities $Ps(Ip_k)$ along every maturity level and that the **more promising options do accumulate on one side** (Fig. 39). This can be used to **reduce the search effort** necessary dramatically.

**S4h)** An **inno-knowledge management** according to the proposed scheme **reduces the search effort** to be better than without it (50:50 inno-game) **exponentially with** its respective **credibility and** the **granularity** (= No. of maturity-levels per search path) of its application (see Fig. 40).

**S4i)** **Any inno-knowledge space can be organized into just 3 dimensions** with a most regular tree-structure automatically fulfilling the subsumption rule. The respective **3 coordinates** are the **domain address $D_i$, the maturity level $M_u$** and the case or **node number $N_k$** (see Fig. 41 or Fig. 42 and (L 17)).

**S4j)** All **inno-knowledge management needs to be based on** a comprehensive and consistent **inno-project management** scheme **and** the respective **supporting IT-systems** and -tools (Rule 14, Fig. 43).

## 5.  THE INTEGRATION OF FINANCE AND INNOVATION MANAGEMENT

In this chapter we will explain how innovation, finance and investment management do fit together. From the point of view of our theory and our model, the inno-gem,

- **innovation management always is finance and investment management too.**

We did so, right from the first and most basic lemmas and definitions (see especially Fig. 1, (D 1) and (L 1) of our theory in chapter 1). We now will try to describe the innovation process from the point of view of an invested Euro, Dollar or which currency so ever. Therefore the decisive question to be answered in this chapter is following one:

> *(Q 15)*  *What is the best budget-allocation strategy for inno-pipes/-projects to minimize their risks and to maximize their profits?*

To answer this question in the following paragraphs, we will most extensively apply and exploit the results, lemmas and definitions of the inno-gem derived and explained in the previous chapters. Therefore this chapter by necessity just has to summarize and apply almost all the aspects of our proposed model (inno-gem) and theory.

Therefor we will assume that any reader of this chapter has read the previous ones and is at least somewhat familiar with our innovation theory. For readers who are jumping into this most interesting chapter, we do provide references to the most important results being used. Even if you think this chapter is easy to understand without having read in detail the lemmas, the definitions and the proofs of the precedent ones, we would very much like to advice any reader really interested in understanding our theory and its respective results to carefully read the precedent ones too. This will help to better understand the important consequences of our innovation theory and our innovation process model, the inno-gem.

## 5.1 Preconditions and basics

In order to be able to come to a consistent and a comprehensive description of the integration of finance and innovation management we would like to first point out the framework and some most basic preconditions of our approach in the following four paragraphs 5.1.1 to 5.1.4.

### 5.1.1 Financial and structural consequences of the innovation definition

As indicated in Fig. 1 and required by Lemma 1 our, better Schumpeter's definition (D 1) of "innovation" already does include the economic success (exploitation) of its underlying inventions and/or know how. On top of that, there is another most important structural consequence of Definition 1 and Lemma 1. This leads us directly to the new Lemma 18:

*Lemma 18 - the comprehensiveness condition of innovation accounting*

*(L 18)   The Discounted Cash Flow (DCF-) curve of any innovation $I_k$ and the corresponding innovation-projects $Ip(I_k)$ and innovation-pipes $IP(I_k)$ must always take into account any cost/effort $Tc(I_k)$ and any benefit/profit $Mp(I_k)$ which can be attributed to $I_k$, regardless in which organization, market and/or location they have been created.*

The importance of this completeness condition for innovation accounting (see alsoFig. 44 to Fig. 46) cannot be overestimated. It requires the respective innovation chain to always be complete. If this is not the case, everything will become much more complicated. This will be discussed at least in some aspects in chapter 5.4 in more details.

---

**Proof of Lemma 18 - inno-accounting comprehensiveness:**
For a single innovation $I_k$ and for individual innovation projects $Ip(I_k)$, (L 18) is nothing but a logical consequence of (D 1) and (L 1). For an innovation pipe IP, being a timely ordered set of individual innovation projects $Ip(I_k)$, (L 14) is also true, cause if (L 14) holds for every project $Ip_k$ within IP, it is also true for the whole set which is IP.   **Q.E.D.**

---

Fig. 44 clearly indicates the structural consequences of Lemma 18. As an example we took a typical DCF-curve for an innovation in the automotive industry. Only if the innovation accounting is performed along the whole value creation chain, preferably within one organization and from start of research until end of sales, the respective innovation accounting does lead to economically sound results. If one is not able to do so, e.g. cause one is not master of the whole innovation chain, the very basis for sound economic and managerial decision making is missing (see additionally (D 5)).

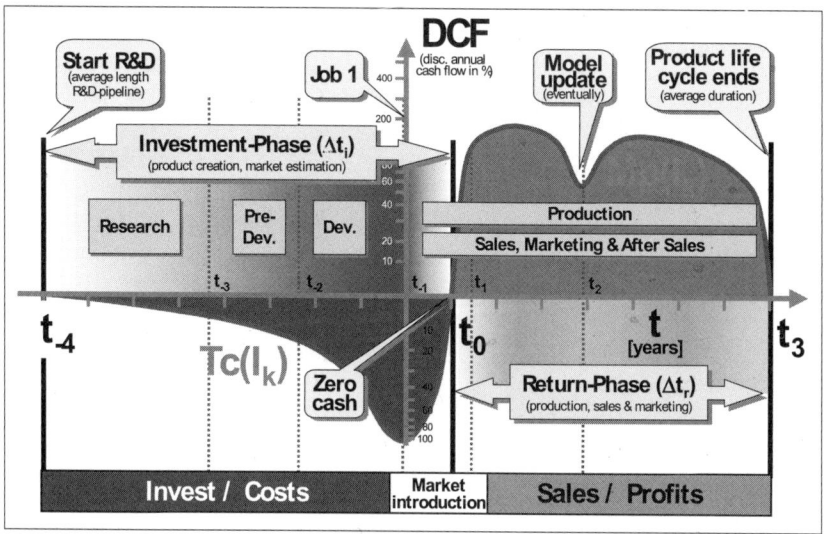

*Fig. 44: Innovation accounting and a typical value creation chain*

### 5.1.2 Innovation phases and the residual value problem

As part of our definition of innovation (see (D 1), Fig. 1 and Fig. 44) we see that every innovation has 2 distinct phase schemes (see Fig. 45) the

- product creation (**Time to Market, TtM** $= t_{-1}-t_{-4}$) and marketing (**Market cycle Time, McT** $= t_3-t_{-1}$) phase scheme and the

- **investment** ($\Delta t_I = t_0-t_{-4}$) and **return phase** ($\Delta t_R = t_3-t_0$) scheme as shown in definition (D 26).

*Definition 26 - the inno-phase schemes*

$$(D\ 26)\quad \begin{aligned} TtM &= t_{-1}-t_{-4} & \sim & \quad \Delta t_I &= t_0-t_{-4} \\ McT &= t_3-t_{-1} & \sim & \quad \Delta t_R &= t_3-t_0 \end{aligned}$$

In general these two phase schemes are fairly identical. The difference is just about 50% of the market introduction phase $t_1-t_{-1}$. Thus, as a first order approximation, we may state that the

- **bulk of the investment** $\Delta t_I$ is done **during** the **product creation (TtM)** and that the

- **bulk of the** corresponding **return/ profit** $\Delta t_R$ occurs **during** the **market cycle time McT.**

In order to be able to integrate innovation-accounting and traditional (balance sheet) accounting, we have to solve a residual value problem too (see Fig. 45). There are two residual values, let's call them the **residual investment A** and the **residual profit B**, we should consider in order to integrate balance sheet and project oriented innovation accounting. During the **investment phase** $\Delta t_I$ we may now rewrite the total investment $Tc_t(I_k)$ into an innovation $I_k$ according to

*Definition 27 - the total innovation investment $Tc_t(I_k)$*

(D 27)  $Tc_t(I_k) = \Sigma disc\_net\_costs\ (I_k) = \int_{t_{-4}}^{t_0} DCF(I_k) + A(I_k) = Tc(I_k) + A(I_k)$

with     $A(I_k) =$ **residual investment**, e.g. from other innovations.

For the return phase $\Delta t_R$ we now write the **total profit $Mp_t(I_k)$ of an innovation $I_k$** according to

*Definition 28 - the total innovation profit $Mp_t(I_k)$*

(D 28) $Mp_t(I_k) = \Sigma disc\_net\_profits\ (I_k) = \int_{t_0}^{t_3} DCF(I_k) + B(I_k) = Mp(I_k) + B(I_k)$

with     $B(I_k) =$ **residual profit of $I_k$** to be used for other purposes.

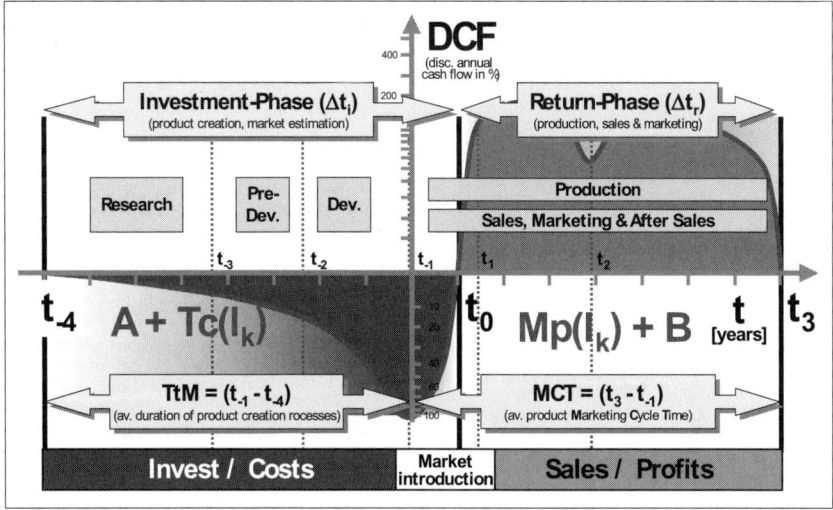

*Fig. 45: Innovation phases and residual values*

We need the parameters A and B for the project- and thus for the phase-oriented innovation accounting to account for (spill over) economic effects between the invest and the return phases of different innovations $I_k$ in an enterprise or in an inno-pipe. Typical examples for such spill-over effects are

- for the **residual investments A** e.g. already **written-off** (book-value =1 Euro) **lab-, shop- or factory-space and/or -equipment** which is still operational.

- for **residual profits B** e.g. already paid **(written-off) resources**, like **stocks, lab-, shop- or factory-space or -equipment** which could **not be sold but still is operational.**

As one might imaging, it is not an exception but rather the general rule, that one starts an innovation with substantial not properly accounted residual investments A and that one ends it with a substantial as well not properly accounted residual profit B, which could be negative too. In order to further facilitate the integration of innovation- and classical accounting, we would like to introduce a further notion, the spill-over value Vsp ($I_k$) of an innovation $I_k$.

*Definition 29 - the spill-over value Vsp($I_k$)*

*(D 29)*    $Vsp(I_k) = B(I_k) - A(I_k)$

The spill-over value Vsp of an innovation can be positive or negative. It accounts for deficiencies of the accounting system in general and for difficulties in appropriately accounting for long term investments in particular. Assuming a decent accounting system and under the precondition of a sufficiently precise innovation accounting system we may assume for any inno-accounting purpose following lemma 15.

*Lemma 19 - the completeness condition of innovation accounting*

*(L 19)   The spill-over value $Vsp(I_k)$ for any sufficiently correct innovation accounting system always is small compared to $Tc(I_k)$ and $Mp(I_k)$. It thus can be neglected, just as the residual investments $A(I_k)$ and the residual profits $B(I_k)$ can be neglected in a first order approximation too.*

---

**Proof of Lemma 19 – completeness condition of innovation accounting:**
   As we saw in definitions of $Tc(I_k)$, $Mp(I_k)$, $A(I_k)$ and $B(I_k)$, any inno-accounting system does sum up and balance as correctly as possible all the costs and benefits of any innovation $I_k$ being accounted. They all are incremental values. Thus an appropriate inno-accounting does financially correctly account for all positive or negative changes any innovation $I_k$ does execute on its environment (the pipe or the enterprise). Thus one always is able, e.g. by a downpayment to or from the respective stakeholders, to make the resulting spill-over value $Vsp(I_k)$ to zero. The same is possible for the respective values of $A(I_k)$ and $B(I_k)$.                **Q.E.D.**

---

Thus respecting the definitions ((D 26) to (D 29)) and the lemmas ((L 18) and (L 19)) of paragraph 5.1.1 and 5.1.2,

- **it is possible to integrate any phase-oriented innovation-accounting into any classical balance sheet accounting system.**

This is a most appealing feature of our proposed new inno-accounting concept, cause a major overhaul of the world wide quite successfully applied balance sheet accounting principles just would not have been feasible, is'n it so?

### 5.1.3 Financing and sequential inno-processing

As a proven consequence of the inno-gem (see chapters 1 and 2.1 to 2.4 especially) we can now state following Lemma 20 regarding the inno-investment pipe.

*Lemma 20 - the innovation investment pipe*

**(L 20)** *Every inno-project $Ip_k$ and any inno-pipe IP always is an investment pipeline too.*

---

**Proof of Lemma 20:**

   As consequence of our definition of an innovation (D 1), an inno-project (D 2) and an inno-pipe (D 3) every inno-project or -pipe starts with a sequence of efforts, better investments which should lead to corresponding benefits, better returns on those investments. This is nothing but what we would like to call an investment pipeline. **Q.E.D.**

---

Once we accept Lemma 20 - which, as illustrated in Fig. 46, we would consider to be pretty evident - it is obvious, that all the optimal design and control rules derived for inno-pipes and the corresponding capacity (investment) deployment strategies do apply for inno-financing purposes too. Thus we would like to sketch the most important consequences of sequential pipeline optimization for inno-financing (see also chapter 2.8, intermediate summary 2) in the following Rule 15.

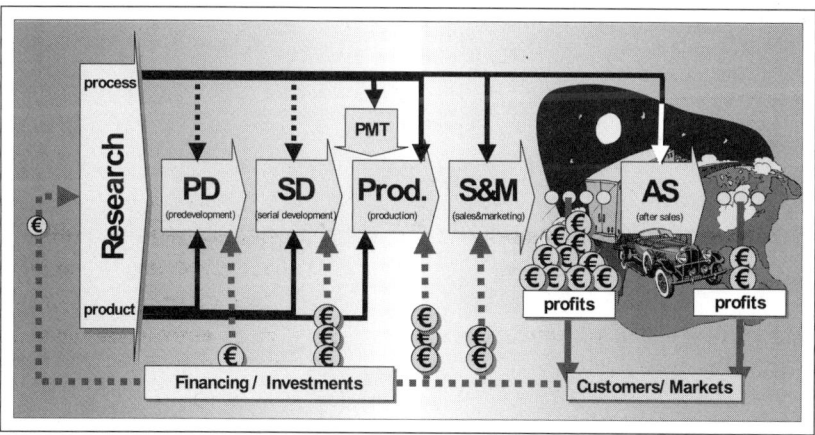

*Fig. 46: Schematic inno-pipe structure and the corresponding financing*

Always remember, there is one real basic optimality condition for any chain or any sequential process to be respected:

- **There may not be a weak link nor a rupture in the chain or sequence!**

For inno-pipes and/or -projects this translates to the following set of rules for a decent inno-financing scheme described below:

---

### Rule 15 - five „golden" rules for inno-financing

**R15.1)** For inno-pipes and -projects in general exponential inno-budgeting strategies are optimal:

**Budget (stage (k+1)) = Budget (stage (k)) * Z**       with $Z \geq 1$
(see (L 10) and Fig. 8)

**R15.2)** In general constant and equidistant „maturity steps" $\Delta Ps$ and „filter rates" **Cp** are optimal:

**$\Delta Ps = Ps(t+\Delta) / Ps(t) = Cp$**     with $Cp \geq 1$ (see (L 11) and Fig. 8)

**R15.3)** There are only 3 basic budgeting/investmentstrategies corresponding to the 3 basic design options for any given inno-pipe:

- **Cascading** $(S_o1)$
- **Buy-in** $(S_o2)$      }    (see Fig. 8)
- **Paralleling** $(S_o3)$

**R15.4)** As a logical consequence of the cost of information principle (L 6) every invested € in an inno-pipe should at any time-step $\Delta t$ bear the same minimal relative failure risk $(1-Ps(t, \Delta t))$. Thus any optimal investment strategy for an inno-pipe IP must satisfy the fundamental inequation (see proof of Lemma 9) for any time-period $\Delta t$:

$$\overline{Ir(IP,\Delta t)} \geq \frac{\overline{C_F(IP,\Delta t)}}{\overline{EVA_S(IP,\Delta t)}} \tag{L 9}$$

with    $\overline{Ir(IP,\Delta t)}$    = average inno-success rate of pipe IP in period $\Delta t$,

         $\overline{C_F(IP,\Delta t)}$    = average failure cost per IP-project and period $\Delta t$ and

         $\overline{EVA_S(IP,\Delta t)}$    = average EVA of an inno-success within IP in period $\Delta t$.

As a one sentence summary, we may state that, especially from a financial point of view,

- **it's the pipe, not the project, who earns the money and who needs optimization!**

---

**R15.5)** Every inno-pipe IP can be interrupted at any stage k without disturbing it, provided that

- **it exists an „intermediate inno-market" accepting/supplying the output/input of that very stage k at the terms the following/previous stage would have done.**

This chain-optimization property is derived and illustrated in the intermediate summary S2 of chapter 2 and illustrated in Fig. 8. If such an „intermediate inno-market" cannot be provided nor simulated, the interrupted inno-pipe in general does not work any more!

This last rule (R15.5) is a very important one, cause it does render us most important hints on what we have to do, once we are confronted with „broken" or „incomplete" innovation and/or investment chains or processes. These problems will be discussed in more detail in the chapters 5.4 and 5.8.

### 5.1.4    From investment to sales, profit etc. for steady state inno-pipes

Our definition of innovation (D 1), inno-projects (D 2), inno-pipes (D 3) and of an inno-success metric (D 4) does allow us to do some quite interesting approximations on the transformation of the invested € (budget) into sales, profits etc. for idealized steady state inno-pipes.

Operated in such a steady state condition an inno-pipe does, according to our inno-gem, produce at its output a stream of market sales/profits as a response to a stream of input investments. Thus we may, as a first order approximation, consider an inno-pipe to be a profit/sales producing machine feeding on investments. This leads us directly to

*Lemma 21 - the invest induced inno-pipe output principle*

*(L 21)*    *Every profit, sale etc. generated by an inno-pipe is induced by a corresponding stream of efforts, investments, etc. into the pipe (see Fig. 46). These 2 streams of money are at least in a steady state operation mode of the pipe strongly correlated.*

---

**Proof of Lemma 21 - invest induced inno-pipe output principle:**

Lemma 21 rather has the character of an axiom than that of a deductible rule or truth. Thus it cannot be proved in a methodologically strict sense. Nevertheless we do consider this principle to be pretty evident, cause it just states the „possibility" of an empirically (statistically) verifiable cause-effect relation between inno-investments (cause) and sales, profits, etc. as their respective effects.

If we were not allowed to assume at least a statistical „causality principle", we might as well refrain from doing innovation management theory at all. There is a lot of empirical evidence, that this is neither true nor reasonable. As just one good counter-example, we would like to show the Mercedes R&D-rates 1982 to 2003 in Fig. 47. We thus would very much like stick to this principle.                                                                    **Q.E.D.**

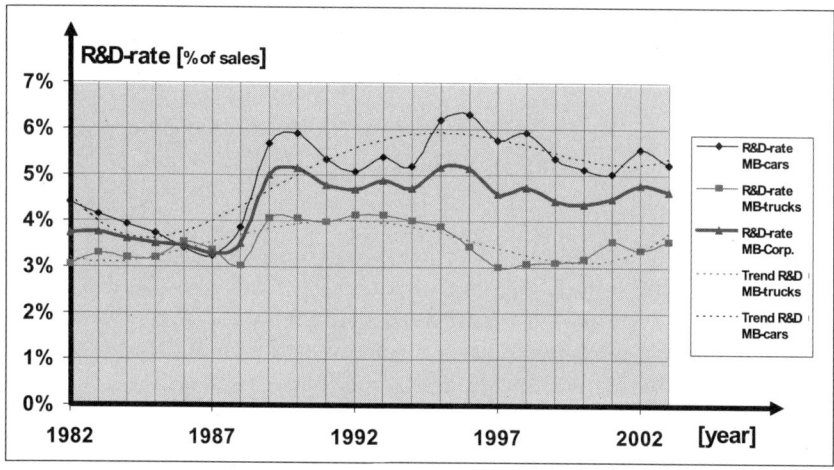

*Fig. 47: The almost constant R&D-rates of the Mercedes car and truck divisions[xxix]*

Applying our inno-gem, this principle and especially the Rule 15 (R15.5) above, first to idealized and later on to real world inno-pipes does render us some quite powerful instruments to do a „model calculation" which performance values and/or figures reasonably well managed inno-pipes could or should have. Some example calculations will be shown in the next chapters (e.g. 5.2 and 5.3).

---

[xxix] Source: Daimler-Benz and DaimlerChrysler annual reports 1982 - 2003

## 5.2 Inno-pipe financing - balance sheet accounting problems

As we have shown in the previous chapter 5.1 on the foundations of inno-accounting, it is a quite reasonable idea to consider an inno-pipe or an enterprise as a timely ordered set of inno-projects. This is even in a formal way true, once the preconditions from chapter 5.1 are respected. Additionally we know that control information and control opportunities for any inno-project and thus for the complete pipe and/or enterprise always go along the time axis (see Fig. 44 and Fig. 45) from the past to the future. Again we do see, that our inno-gem most easily derives some most common, but quite often not too well respected best practices in good inno- and investment management[xxx]. This leads us to one of the most important questions for the integration of finance and innovation management, the question (Q 16) about appropriate inno-control indicators:

*(Q 16) What are essential financial control indicators for good innovation management and how can we build an accounting system rendering this information?*

To answer this question (Q 16), let us first have a closer look on the up to now predominantly applied accounting systems and principles, the „balance sheet accounting principle", as we would like to call it. It forms the basis of the US-GAAP, the German HGB and majority of all the accounting regulations in the world. The major drawback from the point of view of inno- and/or investment-management is, that it does by necessity destroy any causality relation by leveraging return/profit streams with cost/investment streams without taking proper care from where they do come from. This and the resulting inno- and investment-control problem is sketched in Fig. 48 below.

---

[xxx] We would like to remind the reader on our inno-control model (Fig. 5) and the proposed remedies (Fig. 16 and Fig. 17) which in effect do „simulate" the chief engineer function.

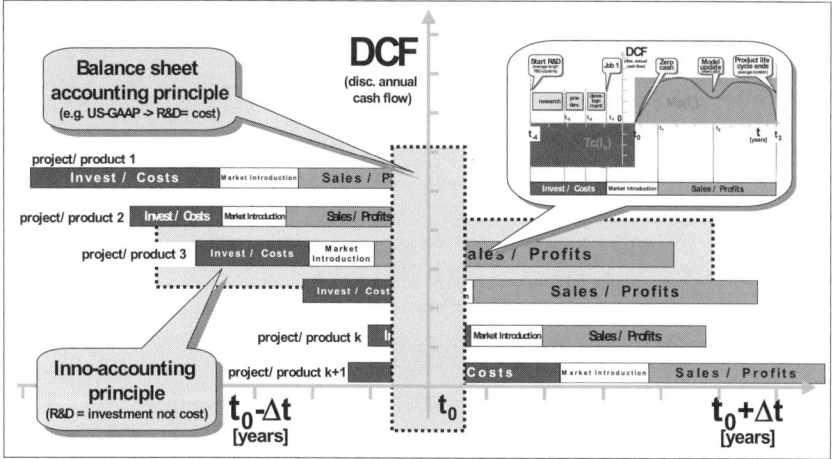

*Fig. 48: Balance sheet and inno-accounting principles*

This mixing (adding up) of costs/investments and returns/profits from different sources (inno- and/or investment-projects) does make any control- and/or success-metric based on these principles (e.g. ROCE, RONA, etc.) completely inappropriate for inno- and/or investment control purposes. Therefor doing our accounting that way, we will never find an answer for (Q 16), cause Lemma 22 below does hold.

*Lemma 22 - the inappropriateness of time-period accounting for inno-control*

**(L 22)** **Any accounting system, which in a given time-period does balance cost/investment and return/profit streams from different sources of an inno-pipe does destroy the financial control information needed for the optimal management and control of that inno-pipe.**

**Proof of Lemma 22:**

There is a formal and informal way to prove (L 22). We think that just looking at Fig. 48 (L 22) becomes evident. In any case we will provide the formal proof for the math-interested reader below:

From (L 18) and (L 19) we know that any inno-pipe is a set of individual inno-projects and that for any time-period we can neglect eventual spill-over values. Thus a financially optimal inno-pipe IP is a set of financially optimal inno-projects $(Ip_1...Ip_N)$. Each innovation project $Ip_k$ is by definition a time-periodic stream of positive (market profits) and negative (investments) DCFs. Any sum of the time-point value of a set of harmonic functions does terminally destroy the phase and the frequency information and thus the characteristic of these very functions (violation of the Nyquist theorem). This corresponds to the destruction of the financial control infos (DCF-curves) for all inno-projects in the pipe and thus for the pipe IP itself. **Q.E.D.**

Obviously classical accounting systems (US-GAAP, HGB, ...) do render us very little help to find an appropriate answer to (Q 16). So we do have to look for new inno-accounting principles (see Fig. 48 and Fig. 51) to tackle that question. We personally are not at all astonished about that. US-GAAP, HGB, etc. implement cash and asset based accounting rules which are designed to ensure that an enterprise does not get illiquid and that it can pay its debts at any time. They are thus designed to guarantee the survival, not the economic success of an enterprise.

Due to that, any inno-accounting must complement and cannot replace classical balance sheet accounting principles and systems. We really do need such a supplement, cause there are even more problems with balance sheet accounting. Once we look a little more into the details of such accounting systems dealing with periodic product projects (see Fig. 48 and Fig. 49), we immediately see that

- **any time-point based (balance sheet) accounting system for a set of periodic product DCF-curves does produce artifacts.**

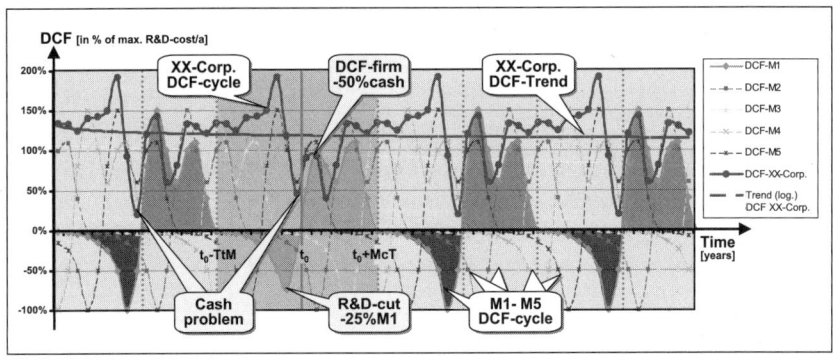

*Fig. 49: Resulting firm DCF-curve for 5-model XX-Corp. with a 25% R&D-cut*

A gentle example of such an artifact is shown in Fig. 49. Here we have assumed our example XX-Corporation to have

- 5 equal product models M1 to M5 at any one time running in parallel.

- operate each model $M_k$ with a 8 year invest (time TtM $\cong \Delta t_I$) comprising of a 4 year research and a 4 year development phase and a 8 year market cycle time McT, including a model refreshment every 4 years after introduction.

- equally distributed each start of a model cycle $M_k$ in time along a (TtM + McT) period of 16 years, which leads to a new modelstart every 4 years.

In this example (see Fig. 49) each DCF-curve of a model is equal and normalized to the maximum negative cash flow occurring during product development (=100%). This is usually at Job 1 (in the automotive industry) at the end of the development cycle. Again, as described in detail in chapter 5.1 (see Fig. 44 and Fig. 45), we distinguish between an invest-phase (until TtM) and a return phase (McT) around an arbitrary zero cash point $t_o$. By the way, the cycle times as well as the DCF-curves for any model $M_k$ are pretty representative for the automotive and corresponding industries. Thus this is by no means an academic exercise and most relevant to quite a few very important industries.

Just looking at the resulting multi-period DCF-curve of XX-Corporation in Fig. 49, we see that its management faces some most severe problems produced by the accounting system and not by economic reality. As most well known in mathematics, the profit function (DCF-curve) of XX-corporation is by necessity periodic, cause it is the sum of the periodic DCF-curves of each model $M_k$. Depending on the periodicity and the shape of these curves, the profit curve of XX-Corporation in Fig. 49 does show by necessity:

- A **main cycle of TtM + McT** duration

- A **constant average annual profit rate** (trend line in Fig. 49) of about **120%** of the maximal investment per year and model. Thus XX-Corporation could finance a model out of its steady state DCF. Again this is pretty realistic, if you look e.g. at a car company having five different models in the market.

- **Constructive (max. 190%) and destructive (min. 20%) interference** and thus large, periodic fluctuations of the firm's DCF of XX-Corporation.

We see in Fig. 49 that these **fluctuations are artefacts**, which do not carry any information on the underlying profitability of XX-Corporation which is by the way constant and positive (120%). Let us suppose now, you are the CFO of XX-Corporation. About a year before $t_0$, when profits were dropping from 190% to just 90% and are heading for a meager 20% you will feel quite uneasy. As in the cycles before, you will most obviously try to cut cost as much as possible, isn't it so? Doing this at $t_0$, you only can try to **cut** model **M1's excessive job1-cost** by, let us assume some moderate **-25%**. As you can see in Fig. 49, now you stopped the profit decline at some 45%, but at what a price. The price to pay is a **terminal loss of some -50%** over the whole M1-product cycle, due to warranty costs and a bad market perform-ance of M1. This is equivalent to loose roughly 42% of a year's profit in one cycle. Obviously **"not taking any action"** is the much **better strategy** for XX-Corporation, isn't it so?

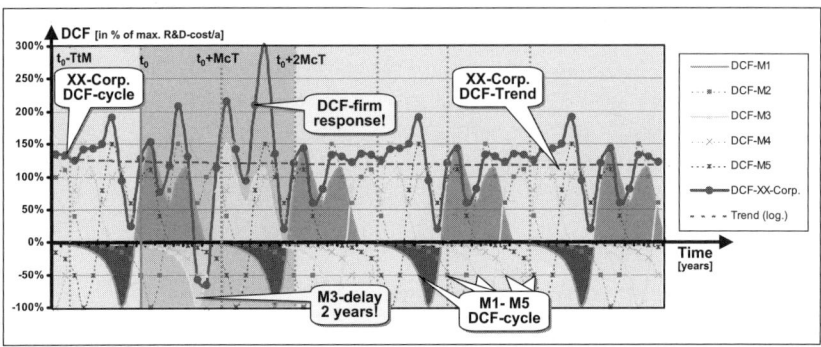

*Fig. 50: Resulting firm DCF-curve for 5-model XX-Corp. with model 3 delayed for 2 years*

Periodic business systems do show even more dramatic effects, than the mild one just discussed. In Fig. 50 we assumed that out of which reason so ever, XX-Corporation has to postpone the launch date of model M3 for just 2 years. As you can see in Fig. 50, the **profit curve** of XX-Corporation in the following market period to McT and in the next one does show some **strong fluctuations** from breathtaking **+300% down to -70%** of a maximal annual R&D-investment. This clearly demonstrates the quite high sensitivity of the DCF-curve of periodic business systems to even minor variations of its com-ponents.

Here again the economic stability and the steady state profitability is not at all questionable. Instead we see a long term positive DCF-rate of 120%. Fig. 50 most impressively shows that any success or control indicator based on classical accounting principles almost by necessity renders misleading indications. **"Ignoring these indicators"** seems to be, at least according to our simulations, **one of the best strategies.** This most clearly demonstrates, that for periodic businesses, **balance sheet accounting principles and indi-cators**

- produce quite **dramatic artifacts due to constructive/destructive in-terference's** and

- **give wrong indications,** which in general do **aggravate the prob-lem/artifact indicated.**

Thus these normal balance sheet oriented accounting principles obviously do not lead us to a suitable answer for question (Q 16) (appropriate inno-control indicators).

### But what would be a better approach for a suitable inno-accounting system?

We do firmly believe, that **the problem**, the disruption of the cause effect (invest/return) relationship, **is already the key to the solution**. So, if we could design an accounting system, which is able to **prevail the business logic** of an individual inno-project $Ip_k$ on a pipe or an enterprise level as well, we would succeed.

This can be achieved, if we start to **look at time rather as a coordinate than as a parameter** of a business system (inno-pipe), just as proposed in chapter 3.5. Doing this, we can, as shown in Fig. 51, generate a kind of a pipe/firm DCF-curve which does prevail the cause-effects relations of all the inno-projects $Ip_k$, which are currently (at time-point $t_0$) in the investigated inno-pipe or firm IP. Naturally we assume all of our proposed **inno-accounting** to be performed **on a discounted and on a total cost basis**.

Provided this and that the spill-over values $Vs_p(Ip_k)$ of the individual inno-projects $Ip_k$ are zero or leveled out, we can be sure that balance sheet and the proposed inno-accounting principle do lead to the same results. This leads us directly to the inno-accounting system design rule described below.

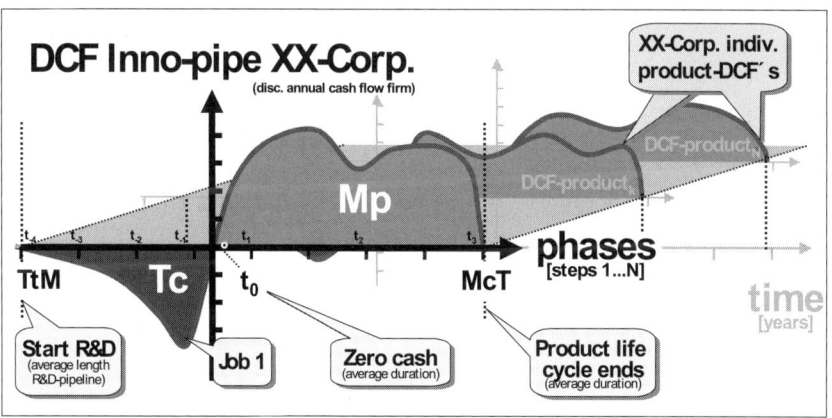

*Fig. 51: Inno-accounting scheme for any firm/pipe (example XX-Corp.)*

*Rule 16 - the 7-step inno-accounting system design rule*

**R16.1)** **Introduce a virtual time-scale of N time steps** around a fixed "zero cash point" $t_0$. This time-scale should cover at least the average TtM and the average McT of your pipe's or companies basic product life cycle.

**R16.2)** **Set $t_0$ to be the qualifying date** of your balance sheet and **recompute all the DCFs** of any project $Ip_k$ active in the pipe/firm with respect to $t_0$ by

- **setting** the zero cash point $t_{0k}$ of each project $Ip_k$ **to $t_0$**
- **dividing** the (real) **time-scale** of each project $Ip_k$ **into** the above **N time-steps** around $t_{0k}$
- **recomputing the DCF-curve** of any project $Ip_k$ in the pipe relative to these above N time-steps.

**R16.3)** **Create the pipe/firm DCF-curve** in each of the N-time-steps around $t_0$ by summing up all the DCF-contributions of each project $Ip_k$ currently in the pipe.

**R16.4)** **Determine an appropriate risk-corridor** by defining appropriate risk-factors for each project $Ip_k$ per year its DCF-value is in the future of the (real time) qualifying date $t_o$.

**R16.5)** **Determine** the **risk-corridor for each project $Ip_k$** in the pipe by multiplying each annual real/planned DCF-value either with 0 for the past or with the risk values of step 4 for the future and recompute the upper and lower boundary DCF-curve for each project.

**R16.6)** **Calculate** the **tolerance corridor** of the pipes/firms **DCF-curve of step 6** by appropriately adding up the respective upper- and lower boundary DCF-curves of the individual projects $Ip_k$ analog to step 2 and 3 as described in step 3.

**R16.7)** Use (D 18) and (D 19) to **set the carry-over values A and B to zero** and compute the "best guess" **RoI-value of the pipe/firm at $t_0$** and its upper and lower boundary according to

*Definition 30 - the inno-pipe return on investments*

$$(D\ 30) \qquad RoI = \frac{Mp(IP) - Tc(IP)}{Tc(IP)} \qquad with$$

$$RoI_{max} = \frac{Mp_{max} - Tc_{min}}{Tc_{min}} \qquad and \qquad RoI_{min} = \frac{Mp_{min} - Tc_{max}}{Tc_{max}}$$

As demonstrated in Fig. 52 by our example simulation of the 5-model XX-corporation with the model 3 delay cash problem, it is not too difficult to follow this 7 step algorithm. It can be highly automated. Additionally it can be easily integrated into existing accounting systems (e.g. SAP R3), once one respects the requirements of our propsed IT- and data-architecture for the inno-phase control system (chapter 3.5). Doing this, it is not too difficult to perform quite elaborate RoI- and investment-sensitivity analyses. This should give the inno-managers in charge a much better insight into the economies of their investments, than traditional tools and accounting principles ever could do.

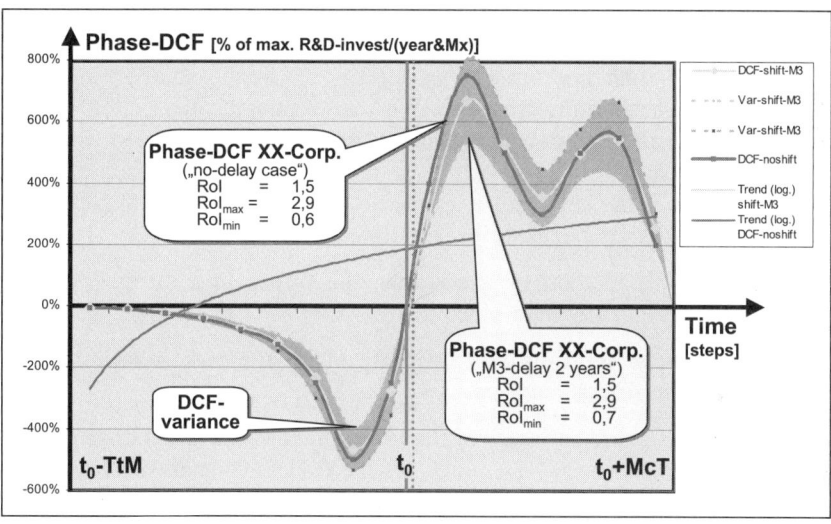

*Fig. 52: Inno-accounting example for 5-model XX-Corp. with model 3 delayed 2 years*

Looking a little more into the details of Fig. 52 we see, quite in contrast to the time-period summary DCF-curve of XX-corporation in Fig. 50, that the 2-year delay of model 3 obviously does only have a minor impact on XX-Corporation´s long term profitability. This is exactly what we would suspect from the DCF-trend line in Fig. 50 and most all from the quite small variations of the DCF-curve and the inno-RoI variations in Fig. 52. This is even more convincing once one considers the quite large tolerance margins assumed (see below) for our DCF-calculations in Fig. 52. They are much smaller than we might expect for we did select quite large variance-values for the DCF-calculations in Fig. 52 of some

| +/- 10% | in the period | $t_0$ | to | $t_0 +2$ years | and |
| +/- 30% | in the period | $t_0 +2$ years | to | $t_0 +4$ years | and |
| +/- 60% | in the period | $t_0 +4$ years | to | $t_0 +6$ years | and |
| +/- 80% | in the period | $> t_0 +6$ years. | | | |

The reasons for that is the fact, that the influence of future risk is limited, cause only projects already begun and in the pipe are considered by the proposed algorithm steps 1 to 7 from Rule 16 to compute the pipe's DCF-curve of XX-corporation and its respective tolerance margins.

Finally, before summarizing the pro's and con's of our new innovation-accounting approach, we would like to remark that, supposed one has an inno-phase control IT-system like the one described in chapter 3.5, one could most easily perform all kinds of model-, project-, factory-, department-, customer-, market- and last but not least time- or phase-sensitivity analyses using our inno-accounting approach. We cannot imagine any better tool nor approach for a most thorough evaluation of any financial inno-investment risk and/or strategy.

To summarize the key features of the proposed inno-accounting approach we would like to mention as its key benefits the following ones:

1) **Best analysis** method for the **status, the future and the past of any investment** into any inno-pipe or firm

2) **Best financial analysis and estimate of all the risk inherent in an inno-pipe** or firm at any time

3) **Best analysis and estimate of the investment value (RoI) of any inno-pipe** or firm we know. This includes the respective analysis of the risks and sensitivities with respect to any of our 4 business coordinates.

4) Only a **limited information acquisition and IT effort** is necessary to profit on all the benefits discribed, provided one has already decided to introduce the proposed inno-phase control scheme (see chapter 3.5).

Naturally there are drawbacks too, once one decides to introduce the proposed inno-accounting system. The most important ones are:

a) The need for a **comprehensive project management system** (tools and processes) as a basis. If it is not already available, one just has to implement it.

b) The need for a **stringent and IT-based accounting system** for projects and cost centers.

c) Some limited (supposed a project-management scheme is available) **extra data acquisition and IT-system and –tool effort.**

We firmly believe that the benefits do outnumber and do outweigh the drawbacks by far. Additionally we see in such an inno-accounting and inno-phase control system a most interesting approach to much better manage the peculiarities of periodic product and business systems (for additional reading see [4]).

## 5.3 Inno-pipe financing - optimal budgeting principles

From Lemma 20 we know, that any inno-pipe is an investment pipe too. Lemma 21 tells us, that in a first order approximation, we may consider any inno-pipe to be a machine transforming budgets into sales, profits etc. as illustrated in Fig. 53.

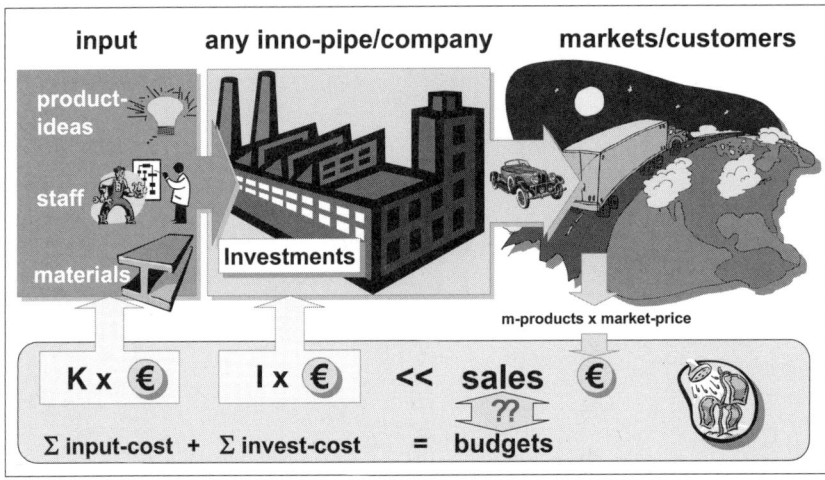

*Fig. 53: An inno-pipe enterprise transforming budget into sales*

From an engineering as well as from a financial management point of view, it is now interesting to get a decent answer on the following question:

> *(Q 17)* **What are the (optimal) transformation rules/parameters from investments/budgets into profits/sales for an inno-pipe/enterprise?**

The inno-gem (see chapter 1), the design and the optimization rules for cascaded inno-pipes (see chapter 2.4, (D 9) to (D 11)) and the financing rules (Rule 15, R15.1 to R15.5) do implicitly give us the answer to that question (Q 17). Thus we may restrict ourselves to only summarize here the most basic rules and to give an example on their respective effects on a real business systems, e.g. our example XX-corporation (see Fig. 54 and Fig. 55). The two most essential rules to be respected for optimal (transformation) inno-pipes are

- the **speed-match condition** and                (see (L 11))

- the **capacity- match condition** .             (see (L 10))

They simply guarantee at any stage and/or time a smooth inno pipe without ruptures and their corresponding inefficiencies and thus guarantee, that any stage or time-period in your pipe does render the same relative contribution to the overall result (sales, profit...). You cannot get better than that, as we showed in the proofs of (L 10) and (L 11). These lemmas do lead us directly to our optimal budgeting principle for steady state inno-pipes and projects (Rule 17).

---

*Rule 17 - the optimal inno-budgeting principles*

In a steady state of any inno-pipe IP the optimal inno-pipe phase-budget $b_k(IP)$ for phase k is a geometric sequence of the form

$$b_1(IP), ...., b_k(IP), ...., b_N(IP), \quad \text{with} \quad b_k(IP) = Z * b_{k-1}(IP) \qquad \text{where}$$

$Z = \textbf{constant} \geq 1$ is equal to the average stage or pipe transfer or success rate (corresponding to an interest rate).

For any inno-project $Ip_j$ and for any inno-project phase budget $b_k(Ip_j)$ for phase k of an inno-pipe of N stages, the optimal phase-budgets $b_k(Ip_j)$ form a geometric sequence of the form

$$b_1(Ip_j), ...., b_k(Ip_j), ..., b_N(Ip_j) \quad \text{with} \quad b_k(Ip_j) = Z * F^{-1} b_{k-1}(Ip_j) \qquad \text{where}$$

$$Tc(Ip_j) = \sum_1^N b_k(Ip_j) \quad \text{is the total Investment of project } Ip_j \qquad \text{and}$$

$F^{-1} = \textbf{constant} \geq 1$ is the average filter factor between two stages of the corresponding N-stage inno-pipe IP.

---

**Discussion / proof of Rule 17:**

We do consider Rule 17 to be just a logical, an economical and even a mathematical consequence of the exponential filter characteristic of any inno-pipe IP.

This does hold under the most evident assumption that there is absolutely no possibility to find neither a-priori "good" investments nor a-priori "better" projects. Then Rule 17 holds under the additional assumption that we do consider complete inno-pipes from small start chances ($Ps_0 \approx 0$) until their very end ($Ps_N \approx 1$) only.

For a more general formula please do follow the definitions and formulas (e.g. (D 9), .., (D 11) and Fig. 9) described in chapter 2.4.        **Q.E.D.**

---

To illustrate, that this principle does give most interesting hints on how to dimension e.g. R&D-departments, we would like to refer to the real world R&D-system example of XX-corporation in Fig. 54 and in Fig. 55.

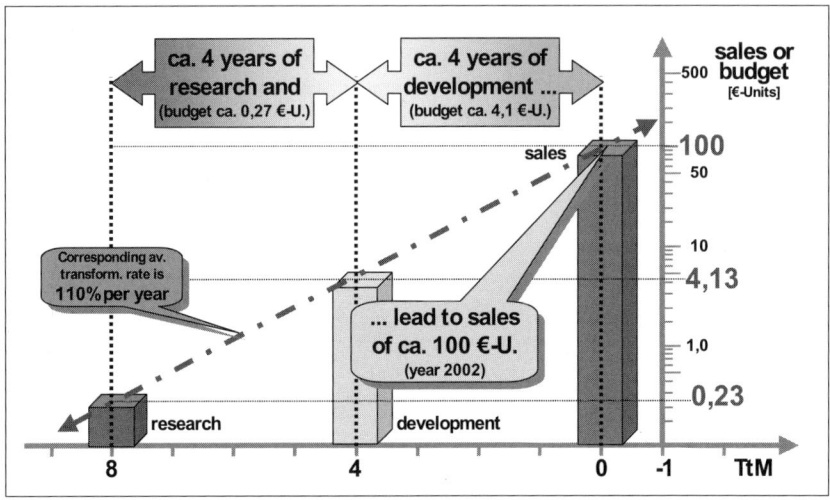

*Fig. 54: The (optimal) budget transformation invest to sales from XX-Corp.*

Obviously XX-corporation worked real hard to achieve a - from the point of view of the inno-gem - quite satisfying status of its 2002 R&D-pipe. It has not been that way for quite some time and there have been several major management efforts to achieve this situation. This does not prove our model nor Rule 17 but it should give them some credibility.

Instead of empirically proving Rule 17, which by the way would be virtually impossible, we are rather interested to discuss this picture using another consequence of our theory and of Rule 17. This is demonstrated in Fig. 55. Here the management of XX-corporation is assumed to try to improve the quite good situation from 2002 (Fig. 54) even further. They did try to do so pursuing a R&D-speed strategy. Thus they shortened the research phase by 2 years. But now the particularities of inno-pipe management (e.g. Rule 17, (L 11) and (L 10)) come to bear and they really do spoil this effort, once not properly executed.

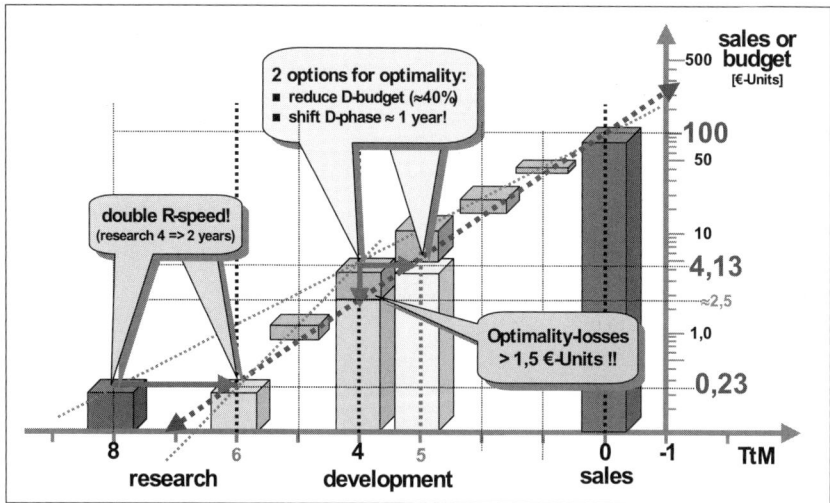

*Fig. 55: An example R-speed strategy and its respective economic consequences*

In Fig. 55 we realistically assume, that the R&D-speed strategy starts with research. It is the smaller department and researchers are believed to be flexible anyway, although this can be doubted considering quite a few real world experiments. This assumed **increase in research speed** (reduction of the research-phase from 4 to 2 years with the output remaining the same) does now **produce a "rupture" in the optimal pipe of 2002** (see Fig. 54). The development department has to digest now the research results coming in at double speed. A double workload is the consequence and thus, if there had not been "quite some room for improvement" in XX-corporation's development, about one half of the research of any year get first piled up and later on probably discarded. Development-capacity is just a limited resource as well. Accepting that, what could be the answer to that problem? Again a short look on our inno-gem does render us **two options out of the dilemma:**

1) **Increase development-speed** by a factor of 2 and thus **reduce development-budget by** about **40%** to achieve the same sales

2) **Reduce** the **development-phase by 1 year** and **increase the research-phase by one year** again to reestablish an optimal R&D-capacity and - speed setting again.

Obviously option 1) is just theoretical one. It will never work for any real development department. Thus we will end with option 2), which tells us that you always have to optimize any inno-pipe as an entity (see the proofs of (L 10) and (L 11) for comparison). Again we would like to remark how nicely our model (inno-gem) and its theoretical consequences fit with what one would consider to be best R&D-management practice.

## 5.4   Inno-pipe financing - dealing with incomplete pipelines

One of the most basic and most important assumptions of our innovation process model (inno-gem) and theory is that

> • **any inno-pipe considered is assumed to be complete from start to the end.**

This assumption forms the basis of our theory and model (see e.g. (D 1) and Fig. 2) as well as the basis of all the conclusions concerning the economic optimization of inno- and/or investment-pipelines. The basic economic problem with incomplete pipes is the simple fact, that in general the return phase ($\Delta t_R$, McT) of the respective investment phase ($\Delta t_I$, TtM) does not work properly for incomplete pipes. This most nasty property of incomplete pipes does lead us to the decisive question (Q 18) to be answered for any incomplete pipe:

---

*(Q 18)   What are suitable options (or strategies) to overcome the return problems, incomplete inno-pipes do have in general?*

---

To answer this question let us look a little more into the details of incomplete pipes or value creation chains and their respective economic optimization problems sketched in Fig. 56. As we can see there, we can restrict the discussion on three basic cases where the economic consequences of the first two cases are already answered, at least implicitly, by the application of our inno-gem. These three cases of incomplete inno-pipes are:

**A) the shortened inno-pipe** - where the **previous stages** of the pipe are **missing** completely.

**B) the unfinished inno-pipe** - where the **return phase** or at least parts of it are **missing.**

**C) the interrupted inno-pipe** - where **parts in between are missing** or not fully developed.

**To case A) - Shortened inno-pipes:**
The nice feature about this case (as well as for case D5-B) is the fact that the underlying economic logic and thus our model (inno-gem) is still valid and applicable. We just have to discuss

• what happens if **parts of the investments $Tc(I_k)$ to come to a certain** market profit $Mp(I_k)$ are **not performed** for an innovation $I_k$?

The answer to that question is simple. Applying Rule 15 (R15.5) straight forward and twofold leads us to the following 2 cases:

---

### Rule 18 - dealing with shortened inno-pipes

**R18.1)** If here is **no "inno-market"** where the missing input know-how can be acquired, one cannot shorten the respective inno-pipe cause there will be **no product and thus no return.**

**R18.2)** If **there is an "inno-market"** where the necessary know how can be acquired, **shortening** an inno-pipe **is always an option.** Its economic success does depend on your respective input know-how costs $Tc(I_k)$ and on the fraction of the "first to market profit" from $Mp(I_k)$ - e.g. due to a potential "speed advantage $\Delta TtM$" relative to the competitors - you can achieve in that market.

---

As a general comment to this case, we would like to state that it represents the normal division of labor between industry and the public (fundamental and applied) research sector. Thus this is nothing new. But the fulfillment of the necessary "inno-market" condition (R18.1 and R18.2) is most often not really and not sufficiently evaluated. A main reason for this is the fact, that this option is advocated by in general technically not too well educated people from marketing, accounting or finance departments. The price to pay is in general a more or less dramatic long-term economic failure from this strategy, although there might have been short-term intermediate economic benefits. Quite a few most prestigious companies died that way!

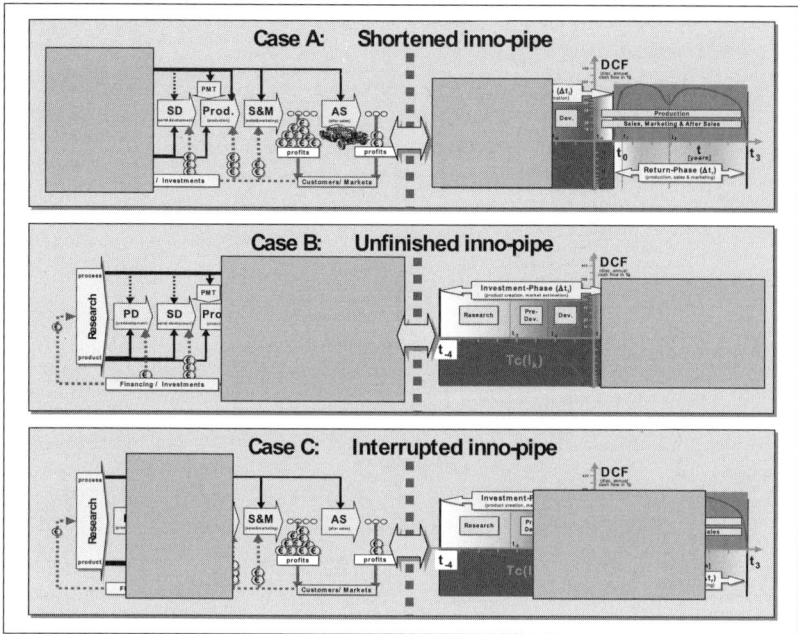

*Fig. 56: Economic effects of incomplete inno-pipes*

## To case B) - Unfinished inno-pipes:

There is not really a need to discuss this case in great detail. Here we assume **that at least parts of the marketing phase** and its respective returns **$M_p(I_k)$ are missing.** Thus this strategy in any case at least diminishes the economic success of an innovation $I_k$ and thus $I_k$ itself (see Definition 1).

Again the availability of a respective "inno-market" (see Rule 15, R15.5), on which one can sell the intermediate know-how generated by the pipe, is the decisive condition. This allows us to state the following two rules:

---

*Rule 19 - dealing with unfinished inno-pipes*

**R19.1)** If there is **no "inno-market"** for the generated results (know-how) of an unfinished inno-pipe $I_k$ there is **no return to cover the costs $Tc(I_k)$** of the investments taken. As a consequence $I_k$ is or will be an economic failure and thus $I_k$ is not an innovation (see Definition 1).

**R19.2)** If **there is an "inno-market"** for the results of an unfinished inno-pipe $I_k$, its respective economic success depends on whether or not one can achieve a market price $\Delta Mp(I_k)$, which is a fair share of the expected total market profit $Mp(I_k)$ of the respective innovation $I_k$ induced by the respective investments $\Delta Tc(I_k)$ already being made.

---

The last rule (R19.2) is or has been the basis of any venture capital and of quite a few investment banking endeavors. There one does trade "expectation values" of future market profits $Mp(I_k)$. History proved that these markets do not work too well. We firmly do believe that the lack of a decent metric/algorithm to evaluate these expectation values is one of or perhaps the main reason for this. We too firmly believe that the inno-gem and our theory presented here is at least a step towards a remedy to this situation.

**To case C) - Interrupted Inno-pipes:**

This kind of inno-pipes is very frequent. We are quite inclined to say that they are almost endemic especially in the assembly industry, which is by far the largest one we have. Thus it really is worth while to have a closer look on the economics of such (interrupted) inno-pipes. For a closer inspection let us assume a one assembler XX-Corporation (red) and a (one) supplier SC (blue) business system, as sketched in following figure Fig. 57:

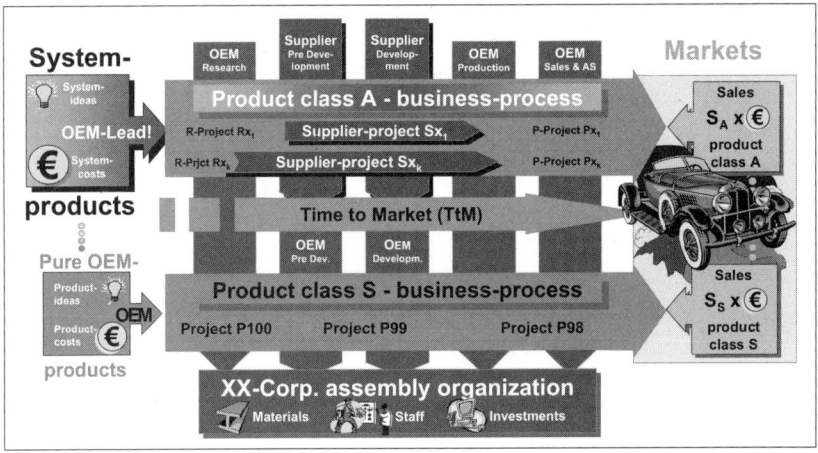

Fig. 57: Example for an interrupted assembly industry inno-pipe structure

Looking at the value creation chain of the system products of XX-Corp. in Fig. 57, we can imagine that its value creation chain (inno-pipe) will have some problems to work properly. This is true from the technical as well from the economical point of view. We will restrict our inspection on the latter one.

As indicated in Fig. 58, XX-Corp. does have a severe economical problem. Being the system-OEM, XX-Corp. is in general the driver of the technical innovations. Thus it has to invest substantial amounts of money (e.g. $Tc_k$(XX-Corp.)) into the development of new system products, but it will only be able to earn a small fraction of the respective component market profits ($Mp_k$(XX-Corp.) in Fig. 58). Thus the expectations to achieve a decent RoI for the new component and the system product are pretty bad for XX-Corp.

Looking at the same problem in Fig. 58 from the supplier point of view, the situation is quite different. As illustrated there, the supplier SC in general has to do about the same investment $Tc$(SC) as the system manufacturer XX-Corp. This is due to the fact that the idea, at least the basic technical specification (research and development phase) and the system integration work has to be performed by the system-OEM XX-Corp. in any case. That leaves only the bulk of the serial development and the start of production investments to the supplier SC. By experience we do think that supplier (SC) and system-OEM investments (XX-Corp.) are fairly comparable in size as sketched in Fig. 58.

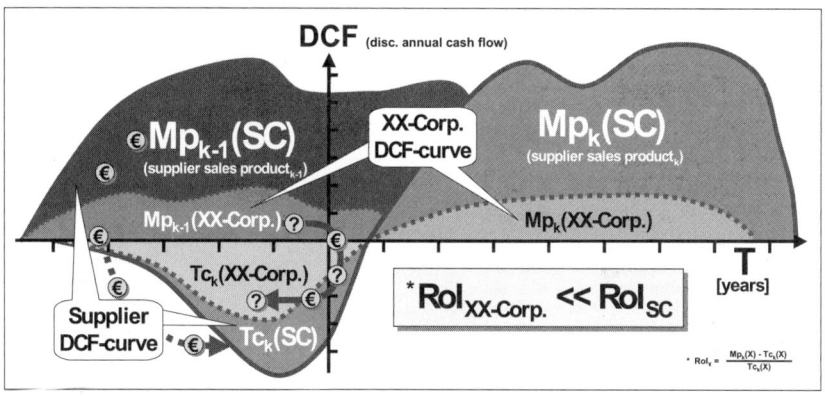

*Fig. 58: Example DCF-curves for interrupted assembly industry inno-pipes*

If we now shift the focus to the marketing phase of our system-component innovation example $I_k$ in Fig. 58, we see that, by economic necessity, the respective market profits $Mp_k$ of the system-OEM XX-Corp. and the one of his supplier SC differ quite a bit. The main reason for this is, that

- the **supplier SC** is always **at the end of the value chain** (production, sales and partially even after-sales) **where profits $Mp_k$ are cashed in.**

- the **supplier always sees economy of scale by selling** the innovative component (e.g. just a few months later) **to all other system-OEMs.**

- in (price-) competitive markets without a predominant "technology leader", the respective **system-OEMs are just forced to integrate the supplier's economy of scale** in their own production- and price-strategies to safeguard their market shares.

At the end of the line of this economic logic, there is a severe and quite **difficult investment / innovation financing problem for the OEM** [xxxi]. This problem is increased, if the financing of the innovation ($Tc_k$(XX-Corp.)) has to be performed from the cash flow $Mp_{k-1}$(XX-Corp) of the predecessor product, the system-component innovation $I_{k-1}$ (see Fig. 58). Having sketched the economic problem of interrupted innovation chains, we should now look for ways out of that dilemma. This does directly lead us to the answer to question (Q 18) for interrupted system-inno-pipes. There are essentially three ways out of the economic problem with system component innovations (see also Fig. 59):

*Rule 20 - dealing with interrupted inno-pipes*

**R20.1)** The **"first-to-market" benefit** of the system OEM is **sufficiently large** to pay for all necessary investments

**R20.2) Make the supplier pay** for all component investments necessary

**R20.3) Form a fair cost/benefit sharing** between supplier and system OEM using all kinds of appropriate real/virtual partnerships and/or joint ventures (see Fig. 59 below).

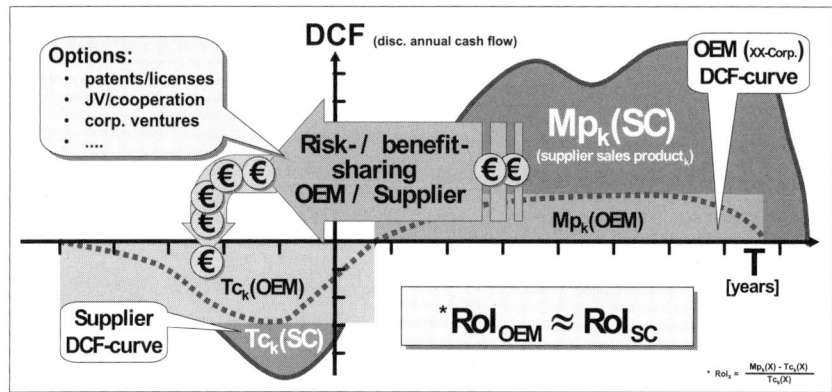

*Fig. 59: Risk- / benefit sharing in real/virtual OEM/supplier partnerships*

[xxxi] A most remarkable example for this problem is the ABS development being financed by Daimler-Benz and Teldix, a small aircraft supplier. Bosch then made the big deal by simply acquiring Teldix (source: Brockhoff, The Dynamics of Innovation, Springer New York 1999).

**To R20.1) First to Market benefit strategy:**

This strategy is a most easy and appealing one for system-OEM management, but it is for sure the **least economic reasonable option.** It is an old fashioned "Premium Strategy" and one needs to be in a highly profitable and relatively unquestioned "technology leader" position to be able to "afford" and to sustain such a strategy for longer terms. Thus it is **not really a reasonable option to deal with interrupted inno-pipes.**

**To R20.2) Make the supplier pay strategy:**

This is at a first glance a quite appealing option for system-OEMs and, if we look at the car industry in particular, there are very strong efforts in the industry to move that way. But nevertheless, there is a most important and most dangerous cloven hoof hidden inside. This cloven hoof is **a massive danger of supplier dependency and loss of technology and market differentiation capability** for the system-OEM. This supplier dependency is a necessary consequence of the differences in the economies of scale between system-OEMs and supplier. Any supplier just has to strive

- to **sell his component** innovations (products) **to all system-OEMs** at equal terms

and thus

- **any supplier is a natural adversary of any system-OEM product differentiation strategy.**

Thus, although easy to manage, this strategy does not work economically. It is only an option for system-OEMs, if

- the **system-OEM** pursuing a "make the supplier pay" strategy does **not** pursue a **"technology leader"** market differentiation strategy as well.

This in turn limits the attractiveness of such a strategy quite a bit and does make it a **viable option only for** companies following a **price-, perhaps a price/quality-strategy** to differentiate from competition. We think these strategies are much more demanding to sustain for a longer time, than any technology leader strategy we could think of.

**To R20.3) Fair cost/benefit sharing:**

The idea and economic logic behind this strategy is sound and simple and its basic idea behind is just to

- **form a real or virtual organization** such that the **economics of a complete inno-pipe are reestablished** in a way comparable to an one enterprise inno-pipe.

This strategy is not quite as easy to implement because

- except patenting/licensing there are **no standardized legal and organizational models** nor blueprints to help to perform this strategy.

Thus management pursuing this strategy really does need courage, creativity and substantial organizational and legal knowledge. On top of that, it will be confronted with a substantial workload, cause quite a few things have to be done form scratch, due to missing blue prints and examples. The trade-off with this strategy is

- **management workload and complexity versus economic potential.**

Following this strategy, management really does have to earn the benefits coming with it, by working very hard. We do not want to go into much detail while explaining the different options to pursue such a strategy. There are quite some articles and there are quite a few benchmark firms around to give an interested reader better and more detailed information on how to carry out such a strategy. Thus we just want to mention that, from our point of view, there are only **three basic options to implement such a strategy:**

1) **Patent-and licensing agreements and policies**

   This is the only legally standardized way how to trade R&D-know how. There is an abundance of expertise and information on the respective pro´s and con´s and there are quite some benchmark companies to learn from. **IBM** is one **of the better candidates** in that respect (see also [11]).

2) **Cooperations and joint-ventures**

   This is a fairly widespread model as well but task, market and company specific. We will thus refrain from further discussing types and modes of such endeavors. We would just like to mention, that these kinds of partnerships are **"bred and butter" management in assembly industries**, although even there, the problem of a fair burden sharing for innovation projects is not really solved. We do hope that our model and theory does help to set up more precise and more reliable metrics and that these metrics help to come to better agreements, on how the burdens and the respective benefits ought to be shared.

3) **Corporate ventures**

   This is an **old method stemming from the industrial revolution inventor-entrepreneur paradigm.** It is becoming more fashionable again these days despite the "high tech venture" euphoria and shock recently (see e.g. [12] for additional reading). The most appealing feature about this approach is that it most easily allows to **integrate industry partners and capital investors to spread out innovations** into new markets. Because again there is an abundance of literature and benchmark examples on this topic, we will refrain from further discussing it. We just want to mention **BT Exact as a benchmark in that respect**. It has installed a most interesting cooperation with NVP-Brightstar, a venture capital firm (see also Fig. 71 and [13]), to company externally capitalize on inventions and innovations generated by their own R&D-/inno-pipe.

## 5.5 Inno-pipe control - from promise to profit (a look inside the pipe)

In the chapter 5.3 we discussed the idea of looking at an inno-pipe as a machine transforming budgets into sales (see Fig. 53 and Fig. 54). In order to give the reader a better understanding how this process works, we now will look a little into the details of such an inno-pipe which

- **transforms the promise of an inno-profit at the start into a really kept one at the end of an inno-pipe.**

In order to be able to do so, we must assume the preconditions and basics of this process and our model described in chapter 5.1 to be accepted and understood by the reader. To structure our walk through the hypothetical inno-pipe of "any-company" presented in the following figures Fig. 60 to Fig. 65), we assume this pipe to be segmented into just 3 stages, as we can see in following figure (Fig. 60):

**1) The investigation phase** where the pipe starts to check the high risk inno-proposals to sort out the most promising ones. In real-life inno-pipes this is a **typical research task**.

**2) The verification phase** where those most promising inno-projects are further developed and tested until only the few technically and economically feasible projects survive. In real world companies this is **done either by pre- or by parts of the serial development departments**

**3) The transfer phase** where **serial development is finalized, production and market introduction** of the most appealing product project **are being prepared.** The technical risk should be fairly low (e.g. Ts>95%) and there should be at least a more than 50%-chance for a decent market acceptance at the end of this phase (e.g. Ms>67%).

During this walk through our example inno-pipe, we will almost see the inno-success function $Ps(Ip_k(t))$ of an inno-project $Ip_k$ or a pipe IP respectively working. As illustrated in the following figures (Fig. 60 to Fig. 65), explicitly computing the success function Ps of a pipe or a project is the best way to control and evaluate it respectively. This is by no means impossible, as some most interesting real world examples do show (see [14] and [15]).

We do believe, any inno-pipe and/or R&D-management should at least give this option a most serious try.

We now do start the example walk through an inno-pipe with the first phase shown in Fig. 61. At this early stage, the start of the investigation phase between $t_{-4}$ and $t_{-3}$ (see also Fig. 3, Fig. 44 and Fig. 45), any inno-project is or should be evaluating which different technology and/or market options are available for the envisaged objective. Cause there are in general quite a few options which look promising, we do expect a "flat" probability density function $Ps(Ip_k)$ as shown in the blue area in Fig. 61. Naturally the center of gravity of this blue area should be on the positive side, because otherwise we would not consider it to be a good project $Ip_k$. In this figure we see another advantage of our Schumpeter like innovation definition (see (D 1) to (D 4)):

- **Measuring** any **inno-project $Ip_k$ by its success $Mp_k$ relative to the respective investments $Tc_k$ necessary, does allow to treat most different inno-projects equally** at least from a financial management point of view.

This in turn does allow us, on the basis of our theory and the inno-gem, to design standardized tools and processes for a better financial management of inno-projects and -pipes.

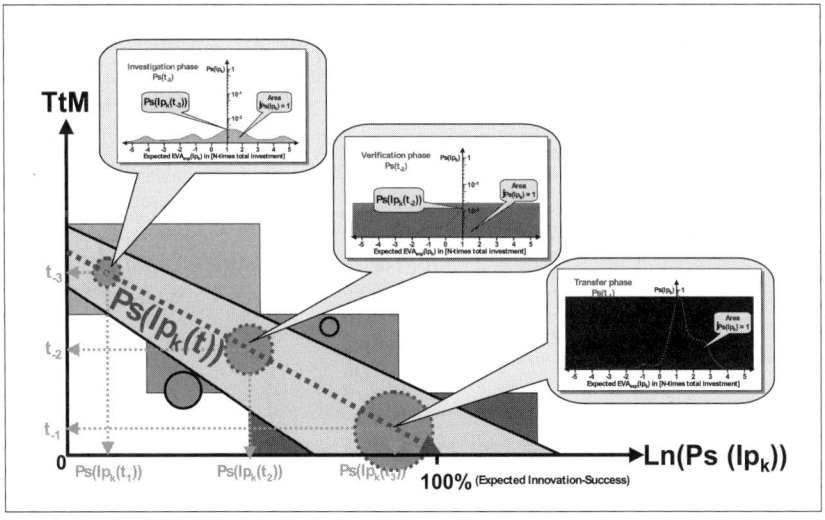

*Fig. 60: Example 3-phase inno-pipe and the respective economic success probabilities*

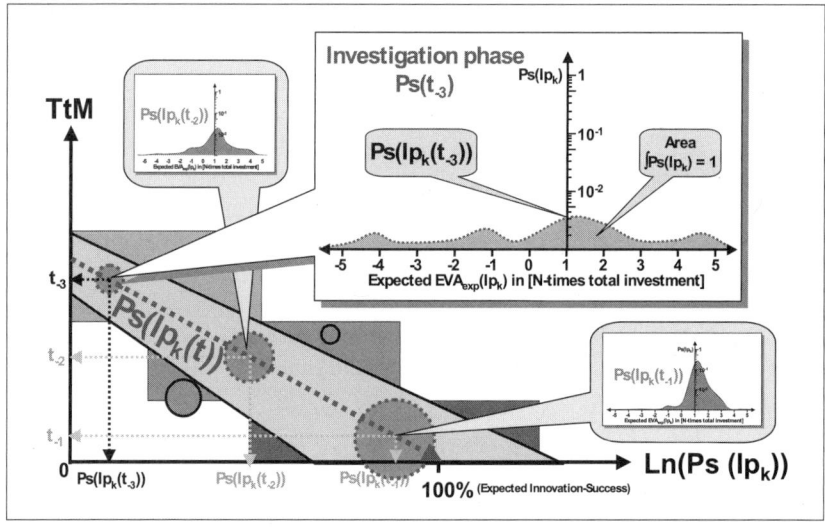

*Fig. 61: Example inno-pipe and the investigation phase economic success probabilities*

Successfully proceeding down the pipe to the verification phase, as shown in following figure (Fig. 62), we now can actually see our "cost of information principle" working (see (L 6), (D 9), (D 11)). Going down the pipe naturally does cost money and effort and thus, it should at least produce some information to help us to successfully continue our endeavor, as it can be seen in Fig. 62. We realize here how, viewed from a financial point of view, a decently working inno-pipe does sort out the "not so good" innovations.

The success **function Ps(Ip$_k$) does become more and more "narrow"**, which is nothing but an equivalent and most precise way to express, that the corresponding realization options for Ip$_k$ become more and more interesting. If the center of gravity of the success-function Ps(Ip$_k$) remains on the right (positive) side and the threshold values for the minimum profitability targets for Ip$_k$ (here it is EVA=1) are met, then we really do have a **promising inno-project.** Thus we really do want Ip$_k$ to be continued at least in this example.

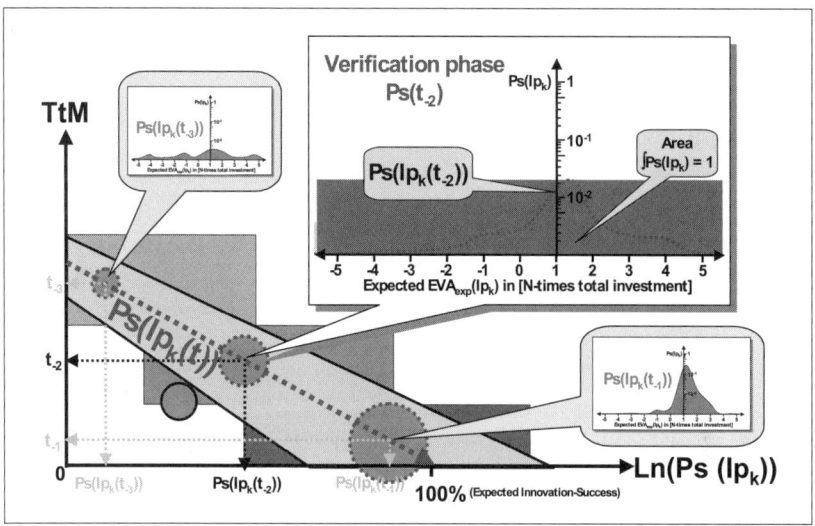

*Fig. 62: Example inno-pipe and the verification phase economic success probabilities*

The results of going further down in our example inno-pipe IP with our (most) promising inno-project $Ip_k$ are shown in the next picture (Fig. 63). Obviously now, in the transfer phase, our success probability **$Ps(Ip_k)$** should at least be greater than some **50%** as mentioned before. On top of that, our "probability density" function Ps should have become much narrower now, showing only a very limited variance. This **variance** directly **corresponds to the up- and down-turn financial success- and loss-possibilities** of the respective project $Ip_k$. This kind of a risk- and risk-sensitivity evaluation is most important for quite large inno-projects, e.g. an envisaged "blockbuster" in the pharma business. With respect to the enormous risks, especially the risk to loose money (e.g. due to law suits or competitor invasions), we do think it is almost mandatory to perform such calculations (see also [14], [15]).

Looking at the same process from the point of view of an individual inno-project $Ip_k$ is shown in the next pictures Fig. 64 and Fig. 65. We do this to illustrate, how an individual project success chance $Ps(Ip_k(ti))$ does fit with the probability success density function Ps in the figures Fig. 60 to Fig. 65. To discuss this connection we assume the inspected project $Ip_k$ to be a "most optimal innovation project" as sketched and discussed in the proof of (L 1) and (L 3) in chapter 1.

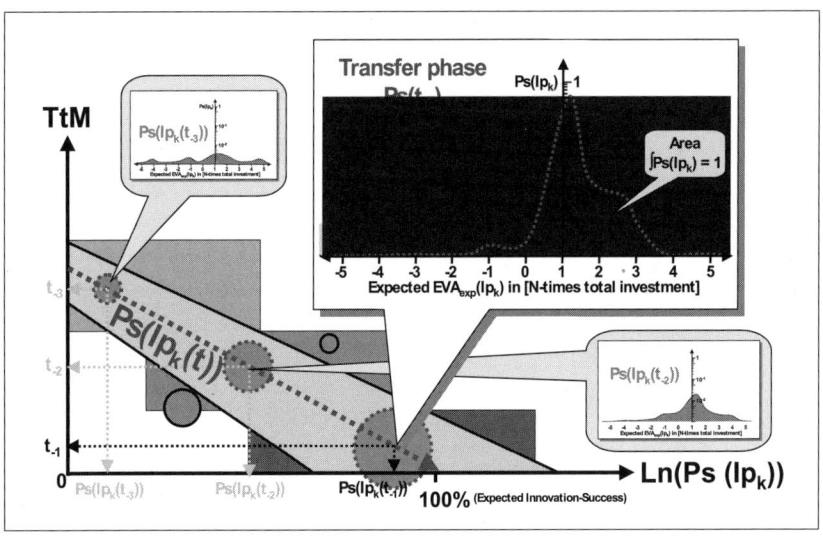

*Fig. 63: Example inno-pipe and the transfer phase economic success probabilities*

The binary **proof tree of** our "most optimal inno-project" **Ip$_k$** does, by nature, **embody all possible outcomes** of this endeavor {**Ip$_{N1}$, ..., Ip$_{NN}$**} and thus this set does **contain** at any stage **all possible values of the success-function Ps(Ip$_k$).** This shows Fig. 64 for the verification-stage of the project.

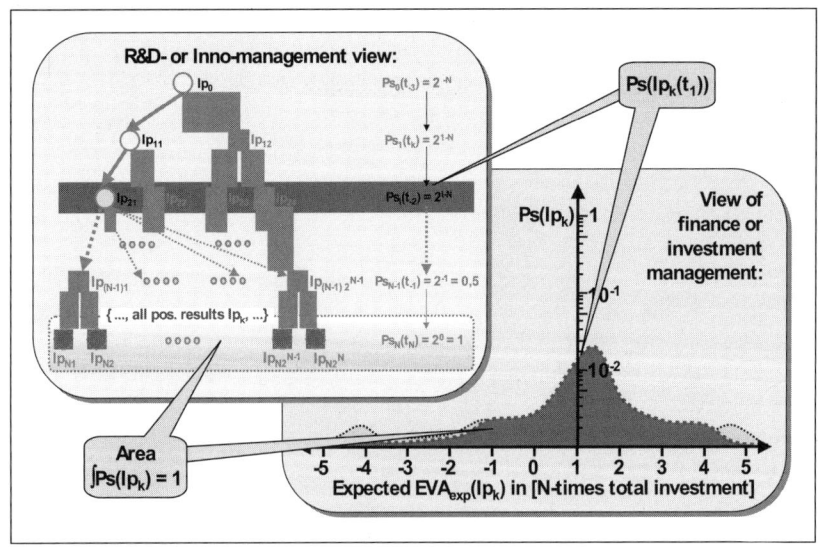

*Fig. 64: An inno-pipe „most optimal proof-tree" with verification phase success probabilities*

Naturally we would see for the start point $Ip_0$ the start density function $Ps(Ip_0)$ shown in Fig. 61 before. Now it is important to understand, that while we proceed down the inno-pipe, even along the most optimal path (yellow circles in Fig. 64 andFig. 65), this very **proof tree remains unchanged.** What does change, while our project $Ip_k$ does proceed along the pipe, are the probabilities to reach the different terminal nodes $Ip_{N1}$ to $Ip_{NN}$. This in turn does represent the change in shape of the probability density function Ps of our project $Ip_k$. Thus, approaching the end of our example inno-pipe (see Fig. 65), the density function just has to get a much more narrow shape. This is even more true, if we approach the end of the pipe on the most optimal path, as assumed in following figure (Fig. 65).

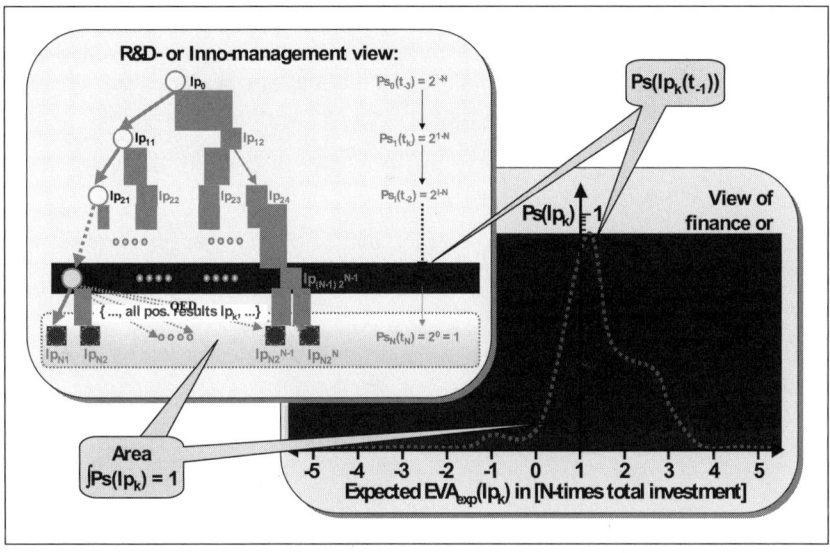

Fig. 65: An inno-pipe „most optimal proof tree" with transfer phase success probabilities

## 5.6 Inno-pipe control - Q-gates to get from promise to profit

There are two main joints who link inno-/R&D- management to finance and investment management. These joints are

- the **phased R&D- and/or inno-budgeting** (see chapters 2.4 and 5.3) and

- the **quality gate and phase control system** (see chapters 3.3 and 3.4)

We now will describe in more detailed way and from an invested budget's point of view, how quality gates do ensure, that the investments into the phases of an inno-pipe IP really do generate money. To do so, we will use again our familiar "must optimal inno-project" introduced in chapter 1 (see proof of (L 1) and (L 2)) to explain, how this linkage sketched in following figure Fig. 66 does work. In this figure you can see most clearly, that

- **any decision rule or gate in an inno-pipe has to be designed from end to front!**

This is most important, cause only at the end of a pipe, there is by definition (see (D 1)) certainty. Approaching the start of a pipe there is hope or expectation at best.

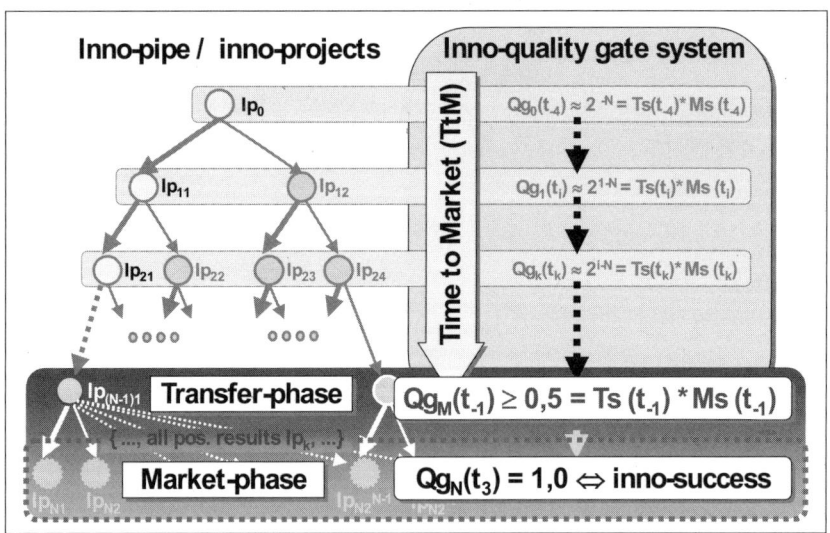

*Fig. 66: Scheme for an inno-quality gate system from start to end (inno-success)*

Assuming our "most optimal" project Ip reaches the market introduction (Job 1) at $t_{-1}$, we would most certainly want

- to have a quite **mature and high quality new product.** Thus we set a remaining 5% failure risk as an acceptable level $\mathbf{Ts_M(Ip)=95\%}$

- to have at least a **2:1 market success chance,** in order not to bet against our own project Ip. Thus we set $\mathbf{Ms_M(Ip)=67\%}$

to be a (minimum) acceptable market success chance for Ip. To make life easier for us, we assume again the most simple case of an incremental innovation Ip with Ps(Ip) =Ts(Ip) * Ms(Ip). With these values we get a success chance Ps(Ip($t_{-1}$))=63% and thus we set

- the **market quality gate** at ($t_{-1}$) to $\mathbf{Qg_M(t_{-1}) = Ps(Ip(t_{-1})) = Ts*Ms = 63\%}$

If we now assume a 10 stage inno-pipe having 3 phases (e.g. investigation-, verification-, and transfer-phase) with equidistant gates and/or project milestones, we can calculate, using our theory, the quality gate thresholds and their respective required technology Ts(Ip) and market success chances Ms(Ip) for the complete optimal inno-pipe. This has been done in following figure for an example of a 10-stage inno-pipe with assumed **technology and market start risks of $Ts_0=20\%$ and $Ms_0=30\%$** respectively.

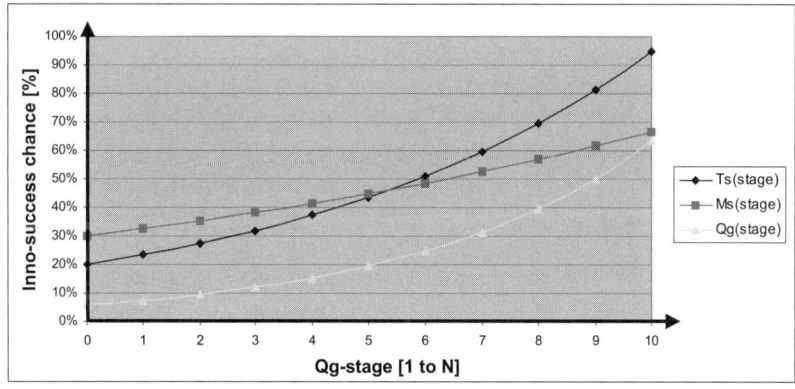

*Fig. 67: Example inno-Qgs and the corresponding Ts-, Ms- and Qg-values*

As we can see in Fig. 67, an optimal inno-pipe - financially as well as from a R&D-management point of view - always does work at a constant **progress rate Cp=27% per stage** or time period towards inno-success. Thus the quality-gate thresholds are a geometric sequence. This is what the Qg-threshold values dictates and the projects $Ip_k$ in the pipe have to comply with. Doing this, R&D-, finance- and investment-management are linked together the best way one can imagine.

Naturally it is much easier to compute these threshold values using our inno-gem and especially the cost of information principle (see (D 8) to (D 11)) than deriving suitable metrics to measure them for real world problems. But now one does have a solid and sound basis to decide on which metric might be suited best for the specific technology, market and company at hand.

## 5.7   Inno-project control - project-risk and project-budget design principles

We do assume that any careful reader of the previous chapters on inno-financing (see especially chapters 5.1 and 5.3) and on inno-pipe design (chapters 2.3 and 2.4) now should have a fair idea on how to organize and budget the projects along an inno-pipe optimally. To summarize the results and the consequences of the inno-gem on inno-project control, we will now discuss the linkage between inno-phases, phase-budgets, quality-gates and project-selection criteria to individual inno-project budgets. To do this we continue to discuss our "most optimal" inno-project /-pipe example from the previous chapters, now on the project-/pipe-level (the microscopic view of the inno-gem), as shown in Fig. 68. For the example inno-pipe shown in Fig. 68 we assume this "most optimal" inno-pipe to have

- start project-success chance **$Ps_0$ > 6%** (Ts>20%, Ms>30%)

- **10 equidistant stages with** an average progress-rate **Cp=27%**

- **3 phases** with 4 Q-Gates **from investigation to market transfer**

- an **end quality-gate with a $Qg_M$=63%** (Ts=95%, Ms=67%)

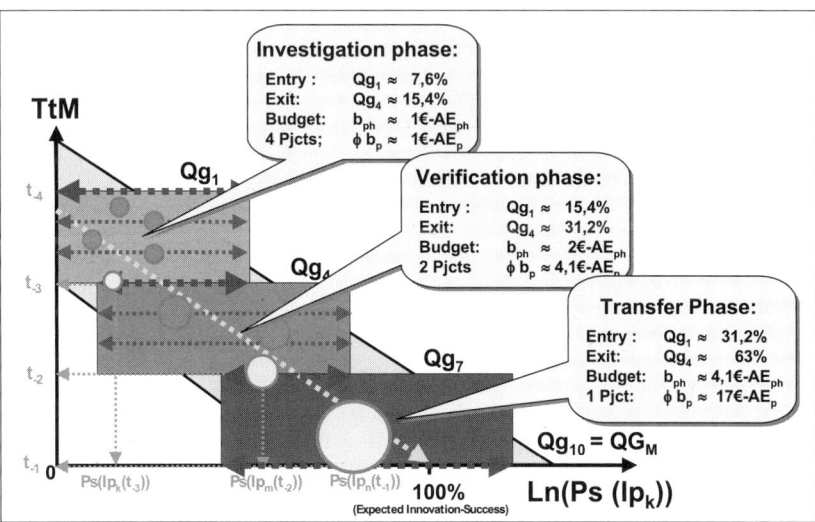

*Fig. 68: Example 3-phase inno-quality gate system with optimal pipe-/project-budgets*

To be able to show how the pipe does handle complexity, we assume that any project is evaluating just one possible project-success alternative. Only the really good alternatives do pass through the gates described in the previous chapter and thus we end with only one fully elaborated (potential) project-success alternative hitting the market at $t_{-1}$.

Such a project-/phase-scheme is directly applicable for simple one innovation products (most common in pharma and in quite a few consumer industries). To apply this scheme to the inno-pipes of the assembly industry (complex product-innovations) as well, we have to make some minor modifications, as illustrated in following figure (Fig. 69).

There we assume a 2 system-innovation pipe with 2 project successes at any one time passing through a five-phase inno-pipe. Thus we have two options, on how we want to look at and control this inno-pipe:

1) **Each numbered rectangle is a small project of just one phase duration** evaluating just one system-inno alternative. These small one-phase projects are evaluated at a quality gate and become fused to some fewer but bigger and more demanding follow-on projects for the next phase until only the fully optimized integrated system-inno project remains in the last step (phase 5).

2) **Each numbered rectangle is a workpackage** of the system project one or two at any phase 1 to 5. At each quality gate the WPs get evaluated and integrated to new ones on a higher system-maturity level. The know-how generated is directly (green arrows) or indirectly (red/blue arrows) used by the follow-on workpackages until at the end the whole system gets compiled in phase 5 and then hits the markets.

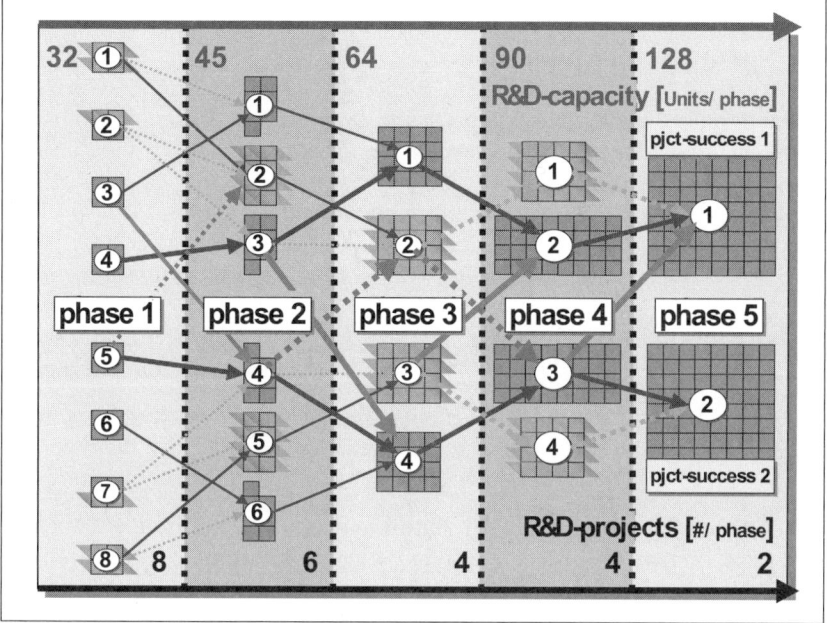

*Fig. 69: 2 Example project-success histories for a gated optimally budgeted inno-pipe*

Just looking at Fig. 69 we see that, from a financial and managerial control point of view, both ways of looking at the pipe render the same results in terms of project control and budgeting. The **main difference** between those two options **is the location where the control is executed.**

- In the first option the **system integration has to be performed by the management of the inno-pipe** himself. This is typical for line-organizations, who are structured according to product models. You will find them quite often in the car industry.

- In the second option the **system integration is the task of the system project management.** Now controlling and control takes place on a partial-project and on a workpackage level.

In both cases, our model and our **budgeting and financial control approach is equally applicable.** What changes is the kind and the granularity level of the necessary accounting (cost-center accounting does not at all work for option 2). Having now reached the end of our discussion on optimal project budgeting principles, we feel pretty confident, that these principles are valid for virtually any kind of inno-pipe and inno-project as well.

## 5.8 Inno-project control - inno-success economy of project- and pipe-exits

In the previous chapter 5.4 we discussed the problems, incomplete pipes do represent for the economic success of inno-projects. This and the fact, that any inno-pipe always is a filter-pipe sorting out a majority of unsuccessful projects, does lead us to Fig. 56 and to following question:

> *(Q 19)*     ***What is the value of unsuccessfully terminated inno-projects?***

A closer look on Fig. 70 tells us, that these supposed unsuccessfully terminated and sorted out inno-projects too carry at least some economic value:

1) They do represent **valuable knowledge on what does not work technically and** to some extent **economically** with respect to the pipe/firm at hand.

2) At least the ones **technically sound and mature,** but from a marketing point of view not suited for the initiating pipe/firm **do carry (potential) economic value too.**

Thus in general the **inno-value creation chain** should not be restricted to the very inno-pipe the respective projects $Ip_k$ are carried out. It **should always consider all the possibilities to generate value,** even outside their own inno-pipes and/or companies.

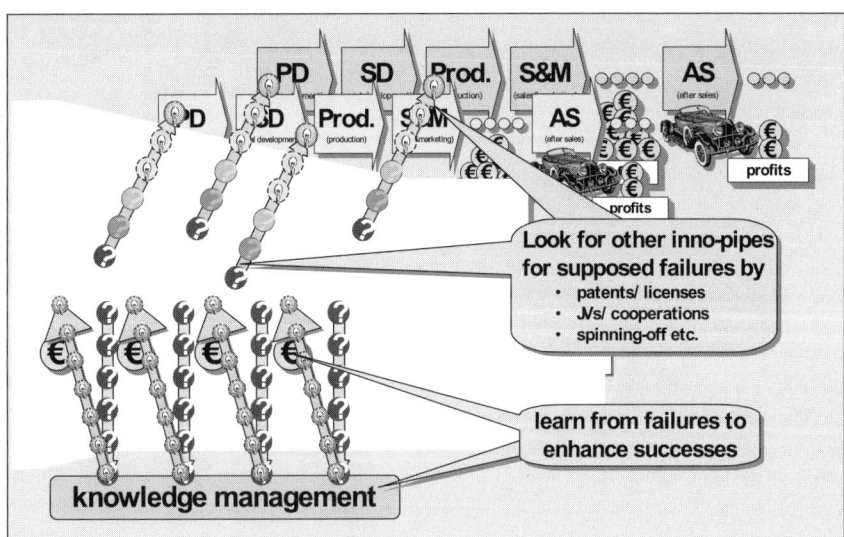

*Fig. 70: Completing the inno-value creation chain*

To complete this value creation process, no big efforts are required in general. The first step is to

- **avoid not using all the value creation options available for any unsuccessfully or successfully terminated inno-project.**

This step in general just needs some managerial effort. In spite of that or perhaps even due to that, this effort is in general not put in or at least quite often not really taken serious. Managerial stubbornness is the only reason we can imagine for this quite suboptimal behavior. If one does undertake a serious endeavor to benefit from the inno-failures as well, there are again two main directions, as described in the following Rule 21:

---

*Rule 21 - external optimization options for inno-economics*

**R21.1)** Make the **best possible use of the knowledge generated** by the inno-failures (see also (L 6)) in your own pipe (see e.g. [5] and chapter 4.3) for knowledge management applications)

**R21.2) Try to complete the value creation chain** for inno-projects you cannot complete successfully in your own pipe or firm by
  - **selling the know-how** and patents obtained so far
  - **forming joint ventures or cooperations** to finalize the innovation
  - **spinning-off** the respective project know-how to some preferably external investor.

---

Both options are most intensively covered in innovation literature. Especially the second option is getting increased attention by large firms, who try to complete their own value creation chain that way.

BT Exact (see Fig. 71 below and [13] for additional reading on how this has been performed) is a quite remarkable example in that respect. For more information on the chances and pitfalls of organized spinning-off and corporate venturing, please read [12].

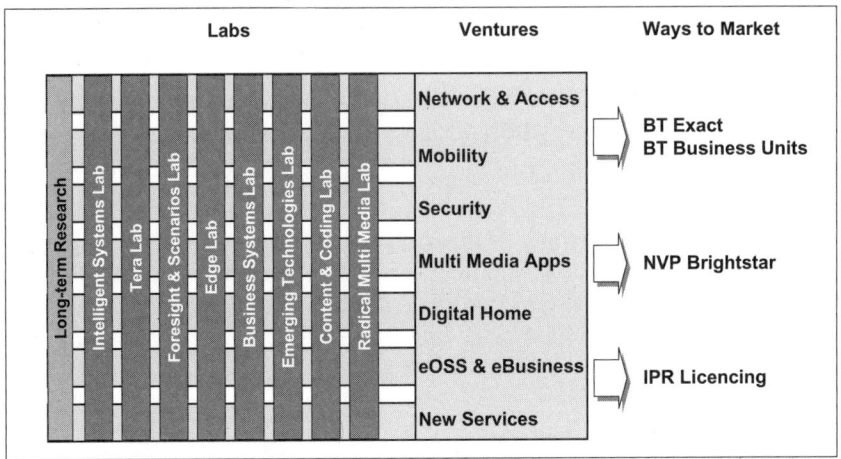

*Fig. 71: The BT Exact multi exit-value creation inno-pipeline design and process set up*

## 5.9 Intermediate summary 5 – integration of finance & inno-management

S5a) **Every inno-pipe is an investment-pipe too** (L 20) and thus there is a **strong correlation** between **pipe-organization, investment-histories and project- and/or department budgeting** (see Fig. 46).

S5b) **Every inno-project has** at least **2 distinct phases in 2 distinct** but **strongly correlated phase-systems,** the **invest ⇔ return** phase system describing the economy of the investment part and the **product-creation ⇔ product-marketing** phase-system describing the logistic and the organizational part of any innovation endeavor (see Fig. 44 and Fig. 45).

S5c) The inno-gem does render **5 "golden rules" to guide the budgeting, financing and controlling** of inno-projects and the respective inno-pipes or R&D-departments as well (see Rule 15).

S5d) **Balance sheet accounting does destroy** necessarily **any business logic** inherent in any inno-project plan or product creation pipeline. **Any success indicator based on these accounting principles** is prone to **produce artifacts rendering misleading information,** which in general does rather **aggravate** than solve **the problem indicated** (see (L 22) and Fig. 48).

S5e) There is a **7-step design** principle for a **new inno-accounting scheme,** which **does prevail the sound business logic** on the project level **for the whole inno-pipe or R&D-organization.** This solves the balance sheet accounting problem for inno-pipes and R&D in general (see Fig. 51 and Rule 16).

S5f) **Any inno-pipe or company can be described** in a first order **approximation as a machine transforming budgets into sales/profits.** These **transformation rules** do follow the "inno-gem investment logic" (see Rule 17) and thus can be **used for rough estimations** for the respective **optimization potentials and pitfalls** of the respective firm or pipe (see chapter 5.3, Fig. 54 and Fig. 55).

**S5g)** The rules derived from **the inno-gem** to calculate the economics of an inno-pipe do **assume always the respective pipe to be complete** from start (idea/plan and investment) to the end of the marketing phase (realization and cash-in). If this precondition is violated, the respective management is strongly advised to **organize their pipe/project** in such a way, that **at least a "virtually complete" inno-pipe is being created** using these 3 approaches/rules for these basic cases:

- **"Shortened inno-pipes"** ⇔ generating a danger of **loosing product-USP´s** (Rule 18)
- **"Interrupted inno-pipes"** ⇔ generating component **inno-financing problems** (Rule 20)
- **"Unfinished inno-pipes"** ⇔ generating **value-exit problems** to solve (Rule 19)

**S5h)** There is an inherent **statistical correlation between** average inno search-tree **complexity,** appropriate **Q-gate criteria and** the respective positive/negative **value exit probabilities** for any decently structured inno-pipe. These in general exponential correlations can be used for the **design of** optimal **Q-gates,** inno-phase **Ps-value requirements** and for optimal selection of **project- and/or inno-phase budgeting** principles for the very inno-pipe being investigated (see chapters 5.5 to 5.8).

# 6. SUMMARY AND BENCHMARKS

We will keep this summary as short and crisp as possible, cause via the rules and the intermediate summaries, there is already quite some compiled information in this book. We organize this summary according to our five chapters one after the other and focus especially on the self-positioning of our model and theory with respect to inno-science and industry standards as we do know them.

**To chapter 1 - the innovation process model (inno-gem):**
Our statistical filter model - the inno-gem - with its three level architecture (macroscopic/filter, mesoscopic/delay-time and macroscopic/proof-tree), the success function $P_s(Ip)$ as the most general metric, its Schumpeter-like innovation definition and finally with the Bayesian calculus rules for the exponential success-function $P_s(Ip)$ is unique in inno-science and in industry as well. It is the first comprehensive, analytical and testable model of the innovation process and we do´nt know of anything comparable (see Fig. 72).

Very much like it is already integrated into our model, keeping the whole inno-pipe in the focus of management is, slowly but steadily becoming a focus in industrial management. There are quite a few firms developing processes and organizational structures to manage complete inno-pipes appropriately, but there is no general model, no organizational and no procedural blue print nor archetype visible up to now. Despite this lack of a solid scientific and methodological foundation, there are quite some very interesting approaches visible in industry. From our point of view, the by far most elaborate and far reaching attempt is the Dupont APEX-research program sketched in Fig. 73. It is an almost classical pipeline and filter design approach and it is in great compliance with our theory. Therefor we recommend it with all our heart for further evaluation and for additional reading (see [16]) as well.

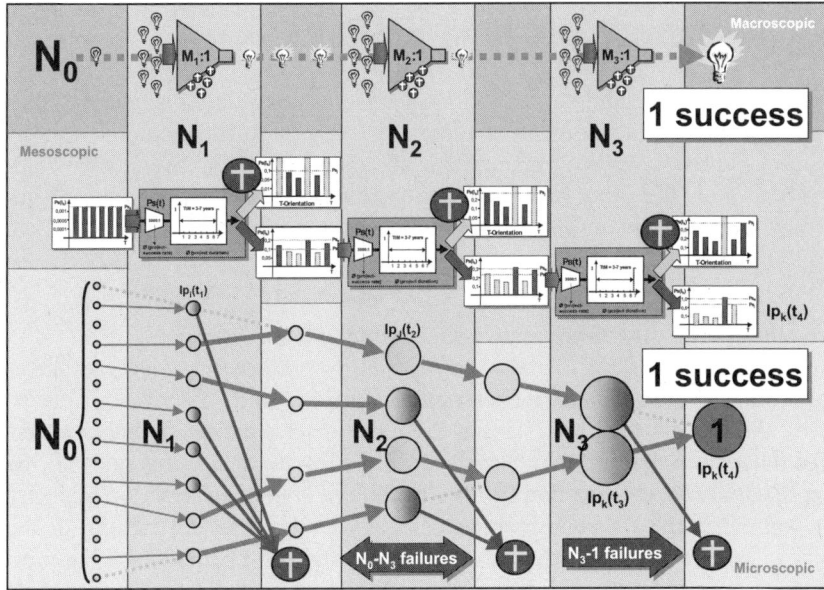

*Fig. 72: The inno-gem - a 3-step idealized ($N_0$ to 1) inno-pipe with all 3 possible views/layers*

## To chapter 2 - the macroscopic view, inno-pipeline design:

Our inno-pipe model, its design rules (Rule 4 and Rule 5), the fundamental inequation (L 9), the capacity and speed match conditions (L 10) and (L 11), our inno-pipe efficiency calculus (D 11) based on the fundamental cost of information principle (L 6), they all are again pretty unique in inno-science and in industry. Our model does for the first time render a sound basis and an analytical model to compute and estimate key economic and non-economic success and design factors of virtually all kinds of inno- and investment pipelines (see e.g. Fig. 9, Fig. 11).

In contrast to a most astonishing lack of appropriate know how in inno-science, there is quite some experience available in industry. For inno-pipe capacity design, there are good examples in the car industry (e.g. Toyota) as well as in a lot of other industries. They all have in common, that they are in general solely based on the experience of some very few managers (chief engineers), cause there is no appropriate general concept nor a blueprint - other than ours - available so far.

On this background the Dupont APEX stage gate and control figures design is even more remarkable (see Fig. 73). It is almost an ideal implementation of what our theory and our model predicts and requires. This is naturally not a proof, but successfully applying such filter designs, especially considering the economic success achieved so far with the Dupont approach (see [16]), makes us feel good with our theory, as you might imagine.

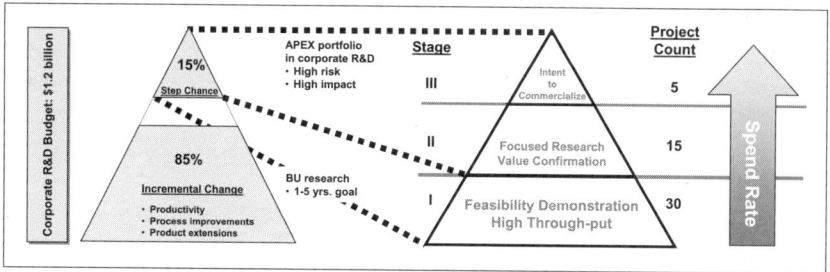

*Fig. 73: The Dupont APEX high risk, high impact research program set up*

## To chapter 3 - Mesoscopic view, inno-pipeline control:

The inno-pipe control model presented (Fig. 5, Fig. 16), the predicted temporal instability of inno-pipes ((L 12), Fig. 15), the phase-control concept (Rule 7) and the proposed portfolio control tool (chapters 3.5 and 3.4) are once more pretty unique and completely new for inno-science and for industry as well.

The proposed integration of our portfolio-approach with the wide-spread and at least partially very sophisticated quality-gate systems ((L 13), Rule 6) in operation in different industries do offer some very interesting possibilities to further improve industry standards in inno-pipe control. The quality of these standards is not so bad after all in industry in general. It is for sure better than that what you find in a lot of scientific articles. According to our theory quality-gate based inno-control systems with equal consideration and evaluation of market (M-) and technology (T-) risk are mandatory.

We do see some very elaborate evaluation schemes for T- and M-risks (see also [14], [15]) at work in industry. A little less elaborate ones are wide spread already and they are becoming industry standard, e.g. in the car industry, where the MDS-system[xxxii] is just one example. The T&M-risk and/or the respective -potential evaluation schemes and processes one can find in industry are in general very sophisticated. Some of the more remarkable examples in industry we do know of are in chemistry Bayer [15] and Dupont [16], in the consumer goods industry Henkel [17] and Kodak [18] and last but not least in the pharma industry Novartis [14], to mention just a few of them.

In order to give the reader some appreciation of the level of sophistication already achieved in some firms, we would like to mention the up- and down-

---

[xxxii] MDS = **M**ercedes **D**evelopment **S**ystem (see also footnote III on page 3)

turn profitability expectation value calculations performed in some companies we do know of (Bayer [15], Novartis [14]). They are performed in a way quite comparable to the figures and sketches we showed in the chapter 5.5 Fig. 60 to Fig. 65). At least we do consider this to be absolutely at par with the highest scientific and professional standards one could imagine. Thus we do hope to see this level of professionalism diffuse more into industry, where ever this may be possible and appropriate. This might not be the case every where, cause these methods do need quite some effort, data and professionalism to be applied successfully.

It is most interesting to notice, that we do not have seen any dedicated time-control scheme at work in industry, neither via portfolios (see chapter 3.5, Fig. 26 to Fig. 30) nor as a deliberately designed and executed chief-engineer control function. Obviously there is still a lack in awareness of this inno-control issue in inno-science[xxxiii] and in industry as well. There are still some very successfully chief-engineers and/or inventor-entrepreneurs (like Gates, Gore, Jobs, ...) on duty these days. They are, by nature, just doing this at least in an informal way. Therefor we cannot understand this lack of awareness at all. This situation is still very much the same as it has been just at beginning of the second industrial revolution with T.A.Edison starting the industrial R&D-lab history in Menlo Park.

### To chapter 4 - the microscopic view, inno-project control:

Our most fundamental proof-tree model (see chapter 1 with Fig. 4, Fig. 6, Fig. 72 and the proofs of (L 2) and (L 3)), the inno-maq (chapter 4.1, Fig. 31) and its respective consequences (Rule 8 to Rule 11) and, most of all, our "value of knowledge-concept" ((L 14) to (L 17)) together with the corresponding rules and the fundamental axioms/preconditions for an inno-knowledge management are at least up to our knowledge completely new. They represent a fundamentally different approach compared to the current state of the art in inno-science as far as we can judge it.

This approach is new to industry as well, but it is not so new and so astonishing, if one takes other sciences e.g. like proof- and control-theory, OR, statistics etc. into consideration. A most appealing effect of our approach is the intimate linkage between the different views (micro-, meso- and macroscopic) and the corresponding levels and advises on how to execute inno-pipe control (e.g. project-management & -control, pipe-portfolio control, pipe-design & -budgeting). This does allow us to derive sound advice, based on the inno-maq (Fig. 31 to Fig. 35) on how to proceed optimally for the just

---

[xxxiii]From our experience, we do not think that there is any awareness about this issue in science at all. On top of that, we found only very limited knowledge about this issue and the respective effects it does produce.

3 basic inno-strategies available for any firm. These advises are Rule 9 for T-push, Rule 10 for D-pull and Rule 11 for incremental innovations. On top of that, for the first time, we can render a formalism and an algorithm to evaluate and thus to optimize the value and the exploitation of knowledge within an inno-pipe or a company.

In that respect we found very little structure and sound advice in inno-science, although there is an overwhelming amount of literature and articles. In industry we also did not see a predominant approach, but we did see quite a few most remarkable examples to learn from. In general, due to the reasons mentioned before, these "benchmarks" are usually not based sound theoretical concepts but on (very) good common thinking and on a solid exploitation of the experience of the people in charge of the respective tasks. This, by the way, demonstrates the "value of knowledge" again.

Incremental innovations are bred and butter technology in industry. The same is true for the more or less stringent quality-gate system approaches applied. Toyota has quite some reputation in the car industry in that respect. A sound T-push driven inno-pipe control with some sophisticated T&M-risk evaluations one can learn from, we did find e.g. at Dupont [16] and at Novartis [14], but there are for sure a lot others too. Additionally we would like to mention Henkel [17] and Kodak [18] with its MAP-mechanism as some most interesting and elaborate examples for sustainable and successful "outside-in" innovations and a D-pull inno-pipe management one can really learn from.

As far as knowledge management is concerned, we do not know of any really sustainably convincing example in industry. This corresponds to missing sound models and theoretical approaches in inno-science. Never the less there are interesting approaches (e.g. the E-BoKs in some car firms) in industry to establish a more formalized KM on top of the good and most effective exploitation of the "biological experience" inside a firm. The problem here is, how to recognize and if, how to transpose the knowledge of these "biological KM-systems" to your own environment. What has been successful in other circumstances and with other people, might not at all work in your own environment. We really do want to see more substantial progress being made in that area in the future. Perhaps this might be even based on the ideas outlined here in the chapters 4.3 to 4.4 and in [5].

**To chapter 5 - the integration of finance and inno-management:**
The total integration of R&D- and investment control via our pipeline model (inno-gem), the calculation rules (see chapter 5.5) for inno-gates and their respective impact on the economies of inno-phases and - searches, the calculation rules for inno-financing (Rule 15) and for inno-project and/or - department budgeting (see chapter 5.7 and Rule 17), the problems of and the

solutions proposed for an appropriate inno-accounting (see chapter 5.2 and Rule 16) are all absolutely unique, at least as far as their consistency and their comprehensiveness is considered. This is true for inno-science and for industry as well.

This most explicitly shows, that our approach, the inno-gem, the statistical filter principle etc. really turn out to be most fruitful, much more that we ever believed it could be. We are looking forward to see this "brake through" successfully pass all verification tests to come. It will then be able and accepted to act as a proven and a solid basis for a substantial improvement in innovation management methods and in inno-science as a whole.

It is a nice side effect of our approach that it too renders the appropriate advice on how to deal with incomplete inno-pipes as well. This case is rather the rule than the exemption, especially for all assembly industries. Thus it is for sure most important. The advises derived from our theory do quite nicely correlate with what is being considered to be best practice in this area (see [12], [13] and Fig. 71). Inno-science does not cover appropriately this very important area. Especially here one can achieve some most substantial economic effects with almost no other resources needed than putting in some managerial brain power. IBM's patent and licensing policy [11] is earning more than 1 Mrd. $, about some 17% of its total R&D-budget that way.

The BT Exact inno-pipe exit scheme shown in the Fig. 71 before is also performing very well. Thus both examples are very strongly recommended to be studied in great detail. With very limited resources, just with some clever organization and management of the pipe and its respective value exits produced, both companies do produce substantial benefits for their share holders. We really do feel quite lucky, that the very approaches taken by these two companies are in complete compliance with the advises derived by applying our theory and our model to such real world problems (see chapters 5.4 and 5.8). Now having finally reached the end of our book, we would like to state, that

- **there are quite some clever inno-management approaches in industry in operation.**

- **inno-science is desperately lacking a comprehensive approach, a consistent model and a much more stringent way of describing the inno-process.**

- **the missing consistent and comprehensive modeling of inno-processes does prevent industry from making more and faster progress towards a more effective inno-management.**

We do hope, that we have been able to contribute at least a little to ameliorate this problem with our approach and our theory described in this book.

This is just a start and for sure not the end of the journey to a more efficient and a more effective inno-management. So let's proceed on this road, cause we personally do believe, the journey already pays for the effort. The odds could be and they have been much worse. Therefor just let's do it – even without the Nike's on.

# 7. ACKNOWLEDGMENT

Two former board members responsible for research and technology (R&T) of a major German car corporation and their respective most fundamental questions to any R&D-system triggered and inspired this work. These questions have been

**"how can we ensure and measure the highest quality
R&D-work possible ?"**

from Prof. Weule and

**"how should an R&D-department (most optimally)
be structured and organized ?"**

from Prof. Vöhringer. Finding suitable answers to these questions has been the start and the motivation of our endeavor to completely describe and model the innovation process and, finally, to compile our results and write this book.

This work would never have been possible without the personal backing, the discussions and the steady organizational support from my former superiors Mr. Hönes and Mr. Knobloch. Most certainly, without that support, we never would have been able to gather, structure, analyze and compile the enormous volume of data and information necessary to come up with these most fundamental solutions we found, our inno-gem, the inno-maq , our rules and our theory.

We would also like to mention here the numerous people disputing with us, putting us persistently on the test by posing more or less nasty question and insisting in appropriate answers. We would like to thank especially the IPM and R&T-Strategy departments of that corporation for providing us with sufficient "rough road testing ground" to be forced to persistently redo and reevaluate our work and to only come up with that most essential and highest quality solution we finally found. We are most certain that without this "rough road", we most probably would have begun too early with implementation work and thus we most likely would have got stuck in "the shallow waters of common innovation management speak".

Aside of the support of numerous colleges from controlling, from other departments and of our personal friends helping, supporting and encouraging us during this multi-year endeavor, we would very much like to mention H.W. Klein, W. Klein and N. Theissen in particular. Together with them we started and formulated the basis our model and theory. The comprehensiveness and the "formal beauty" of its consistency is to a quite large extent their personal credit too.

# 8. LITERATURE AND REFERENCES

[1] J.Alois Schumpeter; "Theorie der wirtschaftlichen Entwicklung", 6th Edition, Berlin 1964, p100f

[2] G.A.Stevens, J. Burley; "3000 Raw Ideas = 1 Commercial Success", IRI 08/95, p16-27

[3] A. Le Corre, G.Mischke; "Projekt Ergebnistransfer", detailed investigation on economic, organisational and non economic R&D success factors of more than 50 R&D-projects (ca. 250 M€ budget, covering a TtM-range of 4-15 years), internal paper 08.07.1998

[4] C.-F. von Braun, "Der Innovationskrieg – Ziele und Grenzen der industriellen Forschung und Entwicklung", Hanser 1994, ISBN-Nr. 3-446-16450-2

[5] W. Klein, "Der Innovationsbeobachter", PhD-Thesis, DaimlerChrysler 2004, tbp.

[6] Ruth Gerrit Huy et. al., benchmark report vehicle development in Japan, DC-internal paper

[7] H-W. Klein, A. Le Corre, G.Mischke, N. Theißen; EPA-Nr. 02007467.0-2221 / 1249781, "Statistische Optimierung eines Prozesses und Computer Programmprodukt zum Durchführen des Verfahrens", 31.03.2002

[8] G.Mischke, PCT/EP03/07447, "Verfahren zur Steuerung, Analyse und Simulation von Forschungs-, Entwicklungs-, Herstellungs- und Vertriebsprozessen", 15.01.2004

[9] Klaus Brockhoff, „Forschungsprojekte und Forschungsprogramme: Ihre Bewertung und Auswahl", F&E Controlling 1973, 3-409-34924-3

[10] T.Ewe, F.Grotelüschen, B. Witthuhn, "Streit um die Wasserstoffwelt" Bild der Wissenschaft Titel, BdW 3/2004, p 84 - 101

[11] E.Ruetsche, Business Development, IBM Corp. Research Lab Zürich, site visit report IBM, "TECTEM - strategic technology benchmarking", TECTEM University St. Gallen, 11/2003

[12] A.Campbell, J.Berkinshaw, A.Morrison, R. van Basten Batenburg, „The Future of Corporate Venturing", MIT Sloan Management Review, Fall 2003, p 30 - 37

[13]   D.G.Brown, Foresight Manager, BT Exact Corp., site visit report BT, "TECTEM - strategic technology benchmarking", TECTEM University St. Gallen, 11/2003

[14]   P.Renner, Strategic Planning & Portfolio Management, Novartis AG, „evaluation of pharma-projects", private communication 11/2003

[15]   J.Genz, Innovation-Engineering Plastics, Bayer MaterialScience AG, „evaluation of R&D-projects at Bayer Polymers", private communication 11/2003

[16]   R.Bingham, CTO, E.I. du Pont de Nemours & Co., site visit report Dupont, "TECTEM - strategic technology benchmarking", TECTEM University St. Gallen, 12/2003

[17]   T.Müller-Kirschbaum, CTO, Henkel KGaA, site visit report Henkel, "TECTEM - strategic technology benchmarking", TECTEM University St. Gallen, 12/2003

[18]   J.C.Stoffel, CTO, Eastman Kodak Corp., site visit report KODAK, "TECTEM - strategic technology benchmarking", TECTEM University St. Gallen, 12/2003

# 9.  SUBJECT INDEX

# 10. FUNCTION AND VARIABLES INDEX

| $I_r(IP,\Delta t)$ | Average innovation rate of IP with respect to a time period $\Delta t$ | *21, 25, 45, 55, 142* |
|---|---|---|
| $K_j(u,v)$ | Knowledge $K_j$ applicable at some stage u in case v | *107, 132* |
| $Ks_l(D_1 \dots )$ | Knowledge space with domains $D_k$ from 1 to n | *127* |
| McT | Market cycle time of product/ innovation $Ip_k$ | *137* |
| $Mp(I_k)$ | Total discounted net lifecycle profit of product/innovation $I_k$ | *2, 5, 136, 139, 151, 162, 166* |
| $Ms(I_k)$ | Market success of innovation $I_k$ | *1, 2, 4-6, 13, 56, 89* |
| $N_C$ | Number of potential customers for a product with $F_C$ and/or $B_C$ | *89, 94, 100, 103* |
| $Ph_k$ | Inno- or maturity phase of IP between 2 quality gates Qg(k-1) and Qg(k) | *63* |
| $P_M$ | Probability of market success for some innovation $I_k$ | *89, 94, 100, 103* |
| Prune   Tree $(I, A_{SC})$ | Basic application algorithm for some negative orientation knowledge in stage/case I under scenario $A_{SC}$ | *114, 116, 132* |
| $P_s(A \mid B)$ | Conditional probability of A under the (assumed) condition B | *13, 56, 89* |
| $Ps(Ip(I_k))$ | Success chance/probability of the innovation project $Ip(I_k)$ | *3-8, 13, 19, 17, 34, 46, 55, 56, 63, 84, 89, 109-11, 124, 171-179* |
| $Ps_0= e^{-\alpha}$ | Entry success chance or start risk for innovation project $Ip(I_k)$ in inno pipe IP | *10, 35* |
| $P_T$ | Probability of technology success | *89, 94, 100, 103* |
| $Q_C$ | Customer accepted quality for $F_C$ | *89, 94, 100, 103* |
| $Qg(Ip,t)$ | Quality gate $\Leftrightarrow$ expert evaluation of the relative success chances $Ps(Ip(I_k))$ at timepoint (maturity) t of project $Ip(I_k)$ within IP | *6, 55, 63, 176-179* |

| RoI | $= (Mp_k - Tc_k)/Tc_k$ return on investment of innovation project $I_k$ | *5, 152, 153, 165, 166* |
|---|---|---|
| $S_p(I_{u,x})$ | Search path (= set of successor nodes) starting from stage/case $I_{u,x}$ | *119* |
| $t_0$ | Zero cash point during market introduction phase of $Ip_k$ (invest cash flow equal return) | *2, 5, 137, 139* |
| $t_1$ | End of market introduction phase of $Ip_k$ | *5, 137, 139* |
| $t_{-1}$ | Start of market introduction phase (job 1) of innovation project $Ip_k$ | *5, 68, 137, 139, 171-174* |
| $t_2$ | Model update for product/innovation $Ip_k$ | *5, 137, 139* |
| $t_{-2}$ | Start of development phase of Innovation project $Ip_k$ | *5, 68, 137, 139, 171-174* |
| $t_3$ | End of the product life cycle of prouct/innovation $Ip_k$ | *2, 5, 137, 139* |
| $t_{-3}$ | Start of pre-development phase of innovation project $Ip_k$ | *5, 68, 137, 139, 171-174* |
| $t_{-4}$ | Start of innovation (project) $\Leftrightarrow$ start of research phase | *5, 68, 137, 139, 171-174* |
| $Tc(I_k)$ | Total discounted net life cycle investment for product/innovation $I_k$ | *2, 5, 136, 139, 151, 162, 166* |
| $Ts(I_k)$ | Technology success of innovation $I_k$ | *1, 2, 4-6, 13, 56, 89* |
| $TtM(t)$ | Time to market (normalized distance in years relative to market introduction of innovation/product $I_k$ | *4, 8, 13, 19* |
| V1 | Business coordinate corresponding to a sink (customers, markets…) | *66* |
| V2 | Business coordinate corresponding to a transaction (process, product …) | *66* |
| V3 | Business coordinate corresponding to a source (lab, supplier …) | *66* |
| V4 | Business coordinate corresponding to a phase/time (date, phase, period …) | *66* |